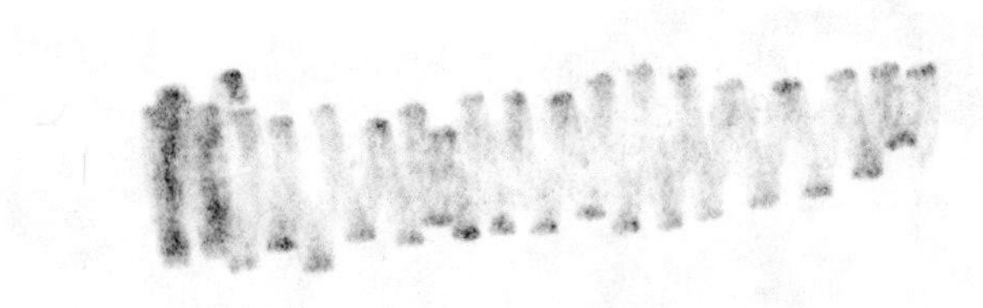

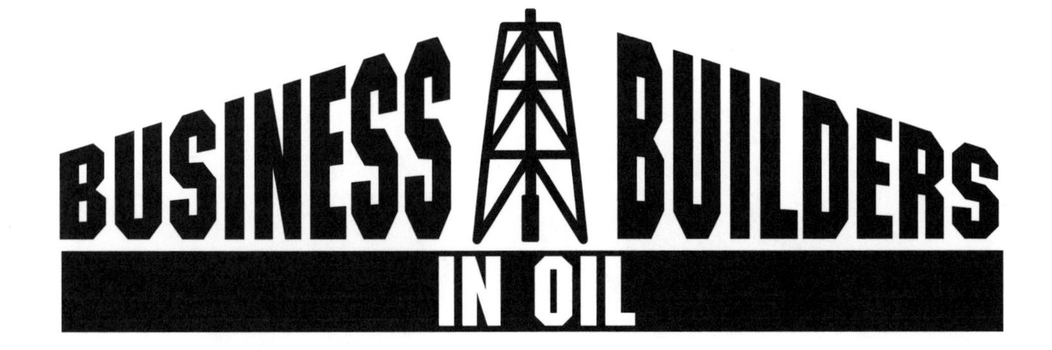
BUSINESS BUILDERS
IN OIL

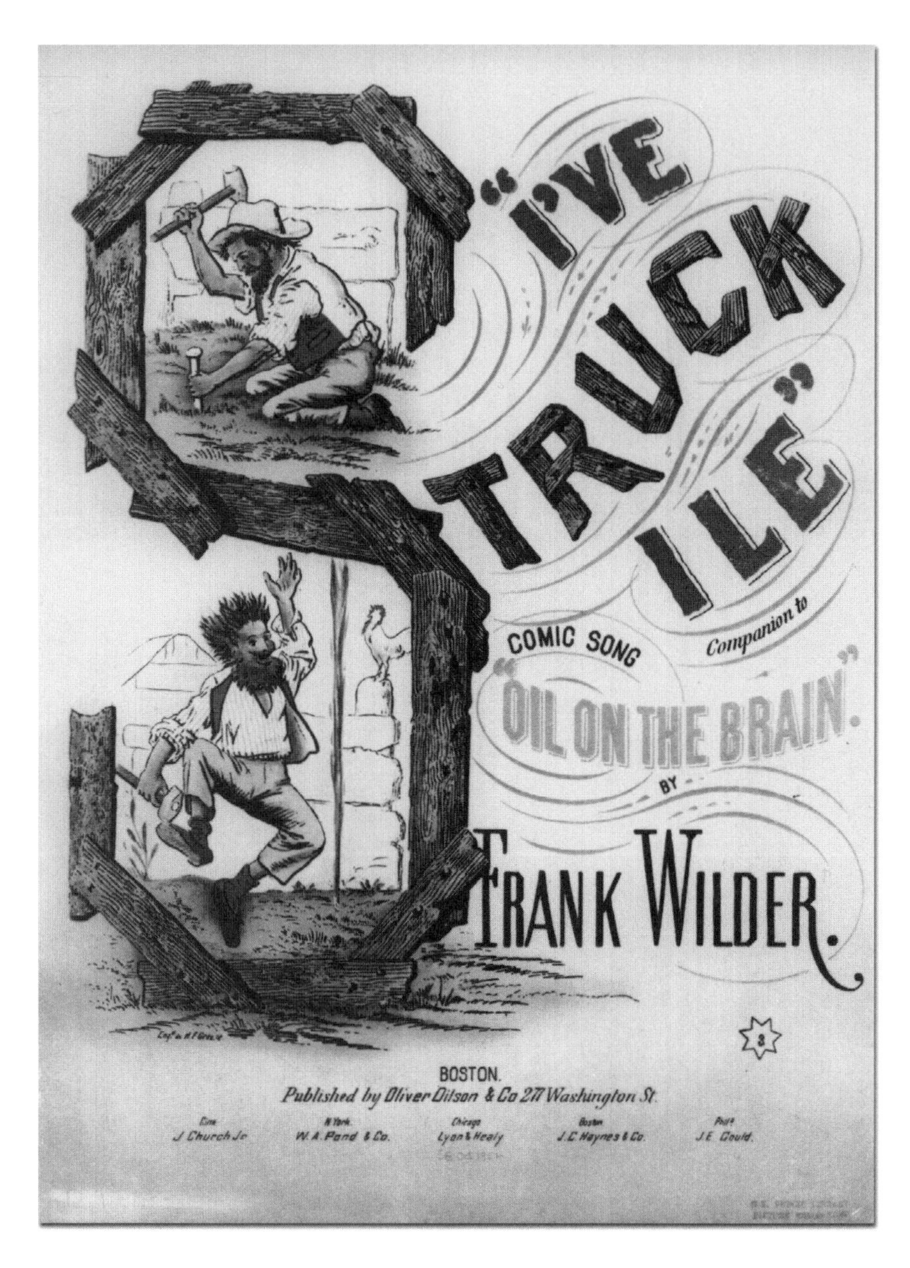

When it was first discovered in large quantities, oil changed daily life in the United States and even became part of popular culture. In the late nineteenth century, people danced to the "American Petroleum Polka" and sang songs like "Oil on the Brain."

Nathan Aaseng

The Oliver Press, Inc.
Minneapolis

To Morris Seydell

The Oliver Press, Inc.
Charlotte Square
5707 West 36th Street
Minneapolis, MN 55416-2510

Library of Congress Cataloging-in-Publication Data
Aaseng, Nathan.
Business Builders in Oil / Nathan Aaseng
p. cm. — (Business Builders)
Includes bibliographical references and index.
Summary: Profiles such pioneers in the oil industry as John D. Rockefeller, Marcus Samuel, William Knox D'Arcy, J. Paul Getty, and Robert O. Anderson.
ISBN 1-881508-56-0 (library binding)
1. Petroleum industry and trade—United States—History—Juvenile literature. 2. Petroleum industry and trade—United States—Biography—Juvenile Literature. [1. Petroleum industry and trade. 2. Businessmen.] I. Title. II. Series.
HD9565.A535 2000
338.2'7282'092273—dc21
[B] 98-050470
CIP
AC

ISBN 1-881508-56-0
Printed in the United States of America

06 05 04 03 02 01 00 8 7 6 5 4 3 2 1

Contents

DRAKE

Introduction

Striking Buried Treasure

In the mid-1850s, New York lawyer George H. Bissell spotted a bottle of oil in a professor's office while on a visit to Dartmouth College. The professor told Bissell that a friend had skimmed the oil from a spring in northwestern Pennsylvania. He wondered if it had any value.

Bissell was aware that oil made from whale blubber commanded a high price as a fuel for lamps. That price was likely to rise even further as the whale population dwindled from overhunting. There was currently no safe, clean-burning, reasonably priced substitute for whale oil. But recently, Canadian geologist Abraham Gesner had developed a process for extracting a product he called "kerosene" from substances such as asphalt and coal. Kerosene showed promise as a new lamp fuel.

Edwin Drake (front right, in top hat) stands before the world's first successful oil well, which he drilled in 1859. The discovery of the Drake Well launched a huge industry with lasting impact on daily life around the world.

People called the Pennsylvania oil "rock oil" because it came from the ground rather than from animals or plants. The word **petroleum**, the scientific name for oil, is a combination of the Greek words for "rock" and "oil."

Human beings have used oil in some form for thousands of years. In ancient times, people used a solid form of petroleum, called pitch, as a waterproofing material. According to the Old Testament, Noah used pitch in building the Ark; the basket in which the infant Moses was found floating was sealed with the same material. The Egyptians covered mummified bodies with pitch, and the Babylonians paved their streets with it. For centuries, a mixture of petroleum and lime called "Greek fire" was a deadly secret weapon. It would ignite when exposed to moisture and was hurled at the enemy with catapults, on the tips of arrows, or in primitive grenades.

Bissell wondered if this "rock oil" from Pennsylvania might prove to be a cheap, plentiful source of kerosene. After discussing the matter with friends, he hired Benjamin Silliman Jr., a prominent chemist from Yale University, to analyze the oil. On April 16, 1855, Silliman reported that rock oil could be refined into lamp fuel as well as lubricants for machines. He was so enthusiastic about the possibilities that when Bissell and his friends formed the Pennsylvania Rock Oil Company, Silliman signed on as an investor. The group then persuaded the New Haven, Connecticut, banker James Townsend to provide financial backing for their venture.

The project was a risky proposition at best. No one knew if rock oil existed in sufficient amounts to be worth pursuing. Even if it did, how did one go about collecting it? The current methods—skimming it off the surface of water or sopping it up in rags—yielded too little oil to be worth the trouble.

The company was still debating how to proceed in the summer of 1856 when Bissell happened to see an advertisement for Kier's Rock Oil Medicine. The ad displayed a derrick, or wooden drilling tower, to illustrate that the oil was found accidentally by miners who were drilling into the earth to find salt. While salt drilling had gone on for centuries and at depths of up to 3,000 feet, few people in the United States had ever considered using the same technique to obtain oil. The very idea that one could mine oil, or perhaps pump it out of the ground like water, sounded far-fetched. Bissell, however, wanted to give it a try.

PETROLEUM, OR ROCK OIL.

A NATURAL REMEDY!

PROCURED FROM A WELL IN ALLEGHENY COUNTY, PA.

Four hundred feet below the Earth's Surface!

PUT UP AND SOLD BY

SAMUEL M. KIER,

CANAL BASIN, SEVENTH STREET, PITTSBURGH, PA.

The healthful balm from Nature's secret spring,
The bloom of health, and life, to man will bring;
As from her depths the magic liquid flows,
To calm our sufferings, and assuage our woes.

George Bissell may have seen an advertisement for Kier's Rock Oil Medicine similar to this one. Samuel Kier claimed that petroleum was "a natural remedy" for everything from rheumatism to blindness.

"Colonel" Edwin Drake

The next step was to get clear legal ownership of the Titusville, Pennsylvania, land on which the oil had been found. For this task, the company turned to Edwin L. Drake, a sickly 38-year-old Vermonter. Although Drake had no outstanding qualifications for this job, fate selected him because of three peculiar circumstances. First, he was staying in the same hotel as Townsend and had impressed the banker with his friendly manner. He also had time to travel because he was on leave from his job as a railroad conductor due to poor health. And third, a fringe benefit of being a conductor was free passage on the railroads, which meant the company did not have to pay his way to Pennsylvania.

Realizing that Drake's credentials were a bit weak, Townsend sent letters ahead to Titusville addressed to "Colonel" Drake. This bogus title had the desired effect. Drake received a warm and respectful welcome when he arrived in December 1857, and he quickly obtained the title to the farmland on which the company would search for oil.

Rock oil was sometimes called "Seneca oil," after the local Native American tribe that had long used petroleum as a medicine.

Drake performed his mission so well that the company, now reorganized as the Seneca Oil Company, assigned him the task of heading up its drilling operations. Drake invested his life's savings in Seneca Oil and then returned to Titusville in May 1858 with his wife, son, and infant daughter. After considerable difficulty, he found an experienced salt driller who agreed to drill for oil. As summer faded into autumn, Drake waited patiently for the man to begin work, but he never showed up. By late autumn, he discovered that the driller, like most other salt drillers, thought Drake was insane. He had no intention of drilling for oil.

Edwin Drake struggled for nearly two years to drill an oil well near Titusville. Locals who believed he would never succeed called the well "Drake's Folly."

Since it would be even more difficult to drill in the winter, Drake had to wait until the following spring to begin work. This time he found a reliable man, William "Uncle Billy" Smith—a blacksmith who made tools for salt drillers—to do the job. The townsfolk of Titusville did not understand what Drake was trying to do, but they were willing to help out the "colonel" in any way they could. Nearly 30 volunteers showed up to help erect the wooden derrick that would support the drill.

Drilling was a crude, slow process. A steam-powered machine dropped a sharp metal rod, called a bit, repeatedly onto the ground so that it pushed a hole deeper and deeper into the earth. The drilling had barely begun when underground water flooded the hole, and the sides of the well collapsed. The workers spent far more time bailing out water than drilling. The project was going nowhere. Drake salvaged the situation by proposing they pound an iron pipe into the ground past the water and then send the drill down through the pipe. This would shield the drill from water and collapsing earth. He ordered 50 feet of pipe, but then suffered another setback when suppliers delivered the wrong size.

Drake's use of pipe to shield the drill was an important innovation that is still used by oil drillers today.

Orders to Close Down

Although Drake assured Seneca's investors that the problems were finally solved, the painfully slow progress discouraged them. One investor traded all his shares in the company for some cigars. The company stopped sending Drake money. Townsend continued to back the project with his own money, but that was not enough. Drake had to rely on the generosity of local merchants to extend him credit so he could feed his family. By August 1859, with the well barely reaching 40 feet, even Townsend gave up. He sent Drake a final payment, with orders to close down operations.

While the letter was in the mail, Drake's pipe arrived at last, and he finally began to make some progress. The rate of drilling increased to three feet per day. On Saturday, August 27, the drill bit had reached a depth of 69 feet when suddenly it slipped into a small crevice. Frustrated, Billy Smith called off work for the day. Drake peered into the empty hole that had once been so promising and saw the end of his foolish dreams. He returned to his family, depressed.

The crew never worked on Sunday, so the well sat unattended until Billy Smith stopped by to check on it. Just as he had dreaded, the hole had filled up with water again. But the liquid seemed darker than usual. Curious, Smith dipped a tin can into the hole and found that the liquid was not water but oil! The discovery came just in time—Townsend's orders instructing Drake to shut down operations arrived the very next day.

A **barrel** is the equivalent of 42 gallons of oil. Today, oil is transported in huge tanks instead of wooden barrels, but the "barrel" still remains the standard unit of measurement for oil.

Drake's well immediately began pumping out 35 barrels of oil per day. With each barrel selling at $40, this represented a sizable fortune. Fired by visions of easy wealth, oil prospectors swarmed to Titusville from all corners of the country in a mad rush to acquire property on which to drill.

Discovery That Changed the World

Drake's oil discovery came at a time when the world was desperate for a new source of lamp fuel and for lubricants to protect the machinery of industry. Kerosene, which was easily derived from rock oil, changed the lifestyles of millions of people. The light it produced extended the working day at factories and offices well beyond dusk and made possible many evening activities in the home. Kerosene became so important to civilization that within 15 years of Drake's successful drilling, it was the United States' largest export among manufactured goods.

As this lamp advertisement demonstrates, the new availability of oil as a source of kerosene gave many more Americans the means to light their homes past nightfall, allowing them the opportunity for leisure activities.

The kerosene era ended when Thomas Edison's electric light bulb came on the market in the 1880s. But by then, rock oil—commonly called petroleum—had taken on new and even more important functions. Not only did petroleum provide light and lubrication, but people also looked to it for heat as oil furnaces began to replace coal furnaces.

A few decades later, a waste product of kerosene production proved to be even more vital to civilization. In the nineteenth century, gasoline had been written off as a dangerous, highly flammable nuisance, to be burned off or dumped in rivers. But by 1905, gasoline had proven to be a far better fuel than steam or electricity for powering the new means of transportation—the automobile. The efficient, cheap power provided by gasoline helped transform the automobile from a luxury toy owned by a few wealthy people to the primary means of transportation for most Americans and hundreds of millions of people throughout the world.

With the rise of gasoline-powered engines of war, oil became crucial to the security of nations. Much of the strategy during World War II, for example, and much of the reason for the Allies' success and Germany's failure in that war, had to do with supplies of oil to fuel the war machinery. As if all this were not enough of a role to play in civilization, oil took on even greater importance with the discovery in the early twentieth century that it could be chemically altered into a new type of material—plastics.

By the time this photograph was taken in 1915, people no longer shouted "Get a horse!" when an automobile passed by.

The Origins of Oil

Most scientists believe that petroleum comes from the remains of tiny plants and animals that lived millions of years ago. These small organisms made their homes beneath the oceans that covered much of the earth's surface at the time. When the creatures died, their bodies sank to the ocean floor and mixed with sediments, small particles of mud and sand.

As the centuries passed, more and more layers of sediments built up beneath the world's oceans. The heat and pressure at these great depths gradually transformed them into sedimentary rock and turned the organic remains within the sediments into a waxy substance called kerogen. As sedimentary rock was buried even deeper, the continued heat and pressure converted kerogen into petroleum and natural gas.

Over time, the oil and gas moved upward through passageways in the rock, pushed by water and other natural forces. Eventually, they reached layers of sandstone and other rocks that were porous, with many small spaces in them. The oil and gas filled these spaces and kept moving until they reached non-porous rocks, which could not be penetrated. Trapped beneath layers of rock, petroleum formed into pools known as reservoirs. As the waters of the ancient oceans receded, dry land appeared over many of these reservoirs, putting them within the reach of humans searching for a new source of energy.

As the world moves into the twenty-first century, oil remains its primary fuel. Oil powers factories that churn out most of the world's products. In many places, the gasoline-powered automobile has made possible an entirely new suburban way of life in which people can make their homes far from the bustle of the industrial cities where they work.

Oil's importance means that demand for it is high and its value, therefore, is high. The oil industry dominates the global market. It is not only one of the largest and richest industries, but its products are essential in the operation of virtually every other industry.

"He Shook the Tree"

To the end of his days, Edwin Drake was convinced that he had personally brought on the oil revolution that so drastically changed the world. "If I had not done it [the drilling], it would not have been done to this day," he insisted.

Although that statement is highly debatable, the fact remains that Edwin Drake was the first person to drill into the earth and pull out a marketable amount of oil. His was the discovery that triggered the high-stakes race to exploit this brand-new product, a race that continues today.

After his oil investments made him rich, George Bissell donated money to Dartmouth College, where he had seen that first bottle of oil. He insisted that the college use his donation to build a gymnasium with six bowling alleys.

In the first flush of a new oil discovery, oil is scarce and, therefore, costly. This makes the land on which it is found extremely valuable. But when the glut of new oil hits the market, oil is no longer scarce. Its price drops dramatically, and so does the value of the land it lies under. Thus, timing and shrewd business sense are crucial in the oil industry. The winners in the oil game acquire undreamed-of wealth, while the losers forfeit their life's savings.

George Bissell was one of the winners. As one of the first on the scene in Titusville, he held the advantage and exploited it. Bissell quickly acquired the rights to a great deal of property in that oil-rich area of Pennsylvania. He made a fortune on the oil that he found and on the skyrocketing prices for which he could resell the land.

Edwin Drake was one of the losers. He was no businessman and had no idea how to take advantage of his discovery. Rather than invest in oil himself, he

bought and sold land for others. He made a good living for a while, until the oil glut dropped prices so low that most of his clients were ruined. By 1866, Drake was broke, gravely ill, and unemployed. His family was surviving on his wife's sewing income.

In this Pennsylvania oil field, forests have been cleared to make room for wooden derricks crowded side by side. By 1865, scenes like this were common throughout the region as fortune hunters strove to "strike it rich" with oil.

Word of Drake's plight eventually reached the people who had profited so greatly from his discovery of oil. The citizens of Titusville raised over $4,800 for him, and the Pennsylvania legislature granted him $1,500 per year until his death in 1880.

One man who knew Drake paid him this tribute: "Honest and upright, he risked his all to develop the oil interest in Pennsylvania, but like many another enterprising man, he shook the boughs for others to gather the fruit." The stories in this book tell how some very shrewd and often ruthless men gathered the fruit shaken loose by Drake and other obscure oil pioneers and transformed it into a giant industry, wealthy beyond belief.

1

John D. Rockefeller

STANDARD OIL: EVIL EMPIRE OR MERCIFUL ANGEL?

The long-dreaded day finally came for the owner of a small Midwestern oil refinery in 1874. As he had expected, the jaws of the giant Standard Oil Company were closing in on him. First, Standard offered to buy the business. When the owner resisted, Standard applied the squeeze, slashing the prices of its oil products to below the cost of production.

The effect of this attack was simple and devastating. The refinery could not sell its products unless it also dropped its prices below cost. Such a move would cause it to lose money. Standard, too, would lose money as long as the price war lasted. But the huge corporation had plenty of cash in reserve. Its directors knew that their small rival would go bankrupt long before Standard felt the pinch. The refiner

In the earliest days of oil discovery, John D. Rockefeller (1839-1937) attempted to unite the entire industry under his control, creating the enormous Standard Oil Company.

had no choice: he had to sell his business before he fell deep into debt.

The independent refiner was enraged by Standard's ruthless tactics. He looked for a way to take revenge on Standard while still getting money out of the business. In the end, he arranged to sell his company to one of Standard's competitors for much less than Standard had offered.

The Standard trap had been sprung. The refiner had no way of knowing that Standard Oil secretly owned the supposed competitor to which he had sold his company. Standard got what it wanted—and at a bargain price.

Such business tactics earned Standard Oil the nation's wrath in the late nineteenth century. For a time, its directors became the very symbol of evil in the world. One newspaper editor wrote, "There never has existed in the United States a corporation as soulless, so grasping, so utterly destitute of the sense of commercial responsibility, and so damaging to the commercial prosperity of the country as is the Standard Oil Company."

The mastermind behind this hated corporation was John D. Rockefeller. The nation so loathed and feared Rockefeller that in the late nineteenth century, the mere mention of his name could scare small children. Ironically, Standard Oil's founder thought of himself as an "angel of mercy" who helped bring prosperity to the nation. A devout man, Rockefeller faithfully taught Sunday school throughout his life, even after he amassed the country's greatest private fortune. He used the wealth he earned from his ruthless business decisions to improve the lives of thousands of people, giving away over half a billion dollars to charity in his lifetime.

Rockefeller gained a reputation as a devil while trying to be a saint because he worked in a wild, untamed new industry with few established rules. His actions reflected an upbringing by parents who had widely conflicting standards of behavior and morality.

Son of Big Bill and Eliza

John Davison Rockefeller was born on July 8, 1839, in rural Richford, New York. His father, William Avery "Big Bill" Rockefeller, was a large, powerful man of questionable character. When John was 10 years old, his father was indicted on charges of rape. Several years later, Big Bill moved his wife, three sons, and two daughters to Ohio and set himself up as a traveling doctor who sold patent medicines and miracle cures. While on the road, Bill secretly married a 20-year-old Ontario woman named Margaret Allen. From that point on, he lived a double life. He eventually left his Ohio family altogether, keeping only occasional contact with them in later years.

"Big Bill" Rockefeller

When he was at home, Big Bill Rockefeller schooled his sons in his cutthroat philosophy of business. He once said, "I cheat my boys every chance I get. I want to make 'em sharp. I trade with the boys and just beat 'em to be sharp traders." Despite Bill's many failings, John admired him and always referred to him as the best of fathers.

Eliza Davison Rockefeller

In contrast to her husband, Eliza Davison Rockefeller was a model of stability and morality. She taught her children to be generous, humble, disciplined, and completely devoted to the Baptist church.

John combined the contradictory traits of his parents. He patterned his personal behavior after his mother. From a very early age, he was abnormally serious, patient, and responsible. By the time he was eight, he was raising and selling turkeys, driving the family's horse and buggy, and handling much of the milking chores. In business dealings, however, he was his father's son, displaying such shrewd business sense that his older sister Lucy commented, "When it's raining porridge you'll find John's dish right side up."

When John was 14, his father drove with him in the horse and buggy the 13 miles into Cleveland, where he left John and his older brother William at a boarding house. For the next year or so, John stayed at the boarding house while attending Central High School. He did not fit in with the rich, well-bred students at the school and spent much of his free time at the Erie Street Baptist Mission Church, which provided him with most of his friends.

While still in his teens, Rockefeller was already acting as a Sunday school teacher, clerk, trustee, and janitor for the Erie Street Church. When he was only 20, he used his business sense to save the church from foreclosure by raising $2,000 to pay its mortgage.

John did not complete high school, but he entered an accounting and bookkeeping program at Folsom Commercial College. Although he was never an outstanding student, Rockefeller had a gift for computing numbers in his head. He finished the three-month program in the summer of 1855 and set out to find a job.

A Toehold in the Business World

Rockefeller targeted firms for which he wanted to work based on their credit ratings. Each morning he dressed in his best black suit and began walking from office to office, asking for a job. Again and again, he left disappointed but not discouraged. He looked for work six days a week for six weeks, often returning to places that had already turned him down.

Finally, on September 26, the commission house of Hewitt and Tuttle offered Rockefeller work at a starting salary of about $4 per week. Hewitt and Tuttle made a living by buying such products as grain and beef in large quantities and then reselling them in smaller amounts. Rockefeller was so thrilled to land the job that he declared the date a personal holiday. Throughout his life, he raised a flag and threw a party on September 26 to celebrate "Job Day."

A member of the Erie Street Church described the young Rockefeller as "not especially attractive," but well liked "because of his goodness, his religious fervor, his earnestness and willingness in the church, and his apparent sincerity and honesty of purpose."

The new 16-year-old clerk took to his job with a passion that bordered on obsession. He showed up for work at 6:30 A.M. and often stayed well into the night. He studied the company's history and finances until he soon knew more about the business than the owners did. Within a year, Tuttle quit the firm, and Rockefeller was promoted to chief bookkeeper, paying and collecting all of the bills.

After two and one-half years, Rockefeller decided his employer was taking advantage of him. While he was performing all the tasks that Tuttle had done, Rockefeller was being paid less than one-third of the salary. He requested a substantial raise, and quit when his boss refused to grant it. Shortly after losing Rockefeller, the company went out of business.

By the time he left Hewitt and Tuttle, 19-year-old Rockefeller had gained enough experience and confidence to go into business on his own. With a loan from his father (who charged him 10 percent interest), Rockefeller joined Maurice B. Clark in opening their own commission house. The partners traded wheat, salt, pork, and fresh produce.

The timing of their business venture was fortunate. Just a few years later, the nation was plunged into civil war. The federal government suddenly needed massive amounts of food to feed its troops, and food was precisely what Rockefeller and Clark had to offer. The company's profits skyrocketed during the four years of war. Rather than spending his earnings, Rockefeller gave 10 percent to charity and invested the rest in land and railroad stock. In 1864, he married Laura Celestia "Cettie" Spelman, a woman he had met when both were students at Central High School.

Although his business partnership prospered, Rockefeller soon felt trapped by the situation. Clark, eight years older than Rockefeller, was a good-natured but overbearing man who treated his partner like a kid brother. Whereas Rockefeller wanted to take risks, such as taking out large loans to build the business, Clark preferred financial security and was content to keep operating on a small scale.

Betting on Oil

While dealing in food, Rockefeller kept his eyes open for new, profitable ventures. The discovery of oil in Pennsylvania in the early 1860s intrigued him. Rockefeller was not interested in drilling for oil—that seemed too big a gamble. But his interest in oil grew when a friend of Clark's, Samuel Andrews, found a way to distill kerosene from oil at a fraction of the cost of extracting it from coal.

When Rockefeller calculated that a gallon of kerosene sold for twice what an entire barrel of crude oil cost, his eyes lit up. He saw that he could make a fortune from a small investment in oil, if only he had a way to transport it to market. Fortunately, Cleveland happened to be located on a main railroad line and was the major shipping center to the west of the Pennsylvania oil fields.

crude oil: oil as it comes out of the ground, before being treated, purified, and turned into finished products in a process called **refining**.

Rockefeller had already begun buying and selling Pennsylvania oil. Now, with Andrews's help, he moved into oil refining—producing and selling oil products such as kerosene.

Oil Refining

The crude oil that gushes from an oil well must be processed before it can be used for fuel or other purposes. To undergo this treatment, petroleum is transported—usually through pipelines or by ocean-going tankers—to an industrial plant called a refinery. Once there, the oil is sent into a large furnace, where heat causes it to boil and turn into a vapor. The vaporized oil then passes into the bottom of a tall cylinder known as a fractioning tower.

As the vapor cools and rises into the tower, it condenses, turning back into a liquid. The different substances, or fractions, that make up petroleum condense at different temperatures. Heavy fuel oils condense at high temperatures, near the bottom of the fractioning tower. Light fractions, like gasoline and kerosene, condense in the middle or top of the tower, where temperatures are lower. At each of these different levels, the condensed liquids are collected in trays and carried from the tower through pipes. The liquids then pass through other processes that use chemicals to remove impurities from refinery products.

Further treatment is often used to produce larger quantities of the most desirable oil fractions. When the automobile became popular, scientists had to figure out how to create enough gasoline to meet the new demand, since gasoline usually makes up only about 10 percent of crude oil. They developed a process called cracking, which uses extreme heat and pressure to turn heavy fractions such as fuel oil into gasoline. Today, about 45 percent of crude oil can be made into gasoline. Another 26 percent is transformed into various fuel oils used for heating and providing energy in factories. Diesel and aviation fuels are other important products of the refining process.

A Standard Oil refinery in Cleveland, built around 1929

Rockefeller kept pouring money into the refining business, and by February 1865, the new Andrews, Clark and Company owned the Excelsior Works, Cleveland's largest oil refinery. But that was not enough for Rockefeller, who continued borrowing money to expand the facility. When Rockefeller took out loans totaling more than $100,000, the cautious Clark objected. As he had done many times in the past, he threatened to dissolve the partnership unless Rockefeller agreed to do things his way.

This time Rockefeller called his bluff. He proposed an auction on the spot, with the company going to the man who bid the highest. His pride at stake, Clark reluctantly went along with the plan. The aggressive Rockefeller outbid Clark and, at the age of 25, assumed control of the company. "I ever point to that day as the beginning of the success I have made in my life," Rockefeller later said.

Four simple principles helped Rockefeller make a fortune in oil refining. First, he strove to increase volume. The more product he could manufacture, the greater his profits. With that thought in mind, Rockefeller expanded and improved his new Excelsior Works refinery and borrowed heavily to build a second, which he called the Standard Works.

Second, Rockefeller recognized the importance of timing. When prices were low and profits plunged, most business owners tightened their belts so they would not fall into debt or even bankruptcy. They cut back production and sold off properties. Rockefeller, however, took the opposite approach. If his rivals needed to sell properties to meet expenses,

When Rockefeller began to invest in oil, his company occupied this second-story office in Cleveland.

that meant Rockefeller could buy these properties at low prices. Therefore, he borrowed and spent huge amounts of money when the oil business was least profitable.

Third, Rockefeller sought efficiency. He made a science out of finding ways to manufacture and transport his products more economically. Rather than pay for oil barrels, the company bought land on which to grow trees and, from these trees, built its own barrels. It also purchased boats, tank cars, and warehouses and drove hard bargains with the railroads for transporting its products. Rockefeller refused to waste money on fancy trappings. Even when he was worth hundreds of millions of dollars, he shared an unpretentious office with another company official.

Finally, Rockefeller pumped most of his profits back into the business rather than spending them. This soon allowed him to make expensive purchases without having to borrow from banks.

Rockefeller was also an expert judge of business talent. He brought in top executives such as Henry Flagler to help him run his ever-expanding company. Rockefeller considered these early days some of the best of his life. "We were all boys together, having a lot of fun as we worked hard every day," he remembered. In 1870, they named the company Standard Oil to indicate a standard of quality that the customer could always expect.

Standard Oil hit only one major speed bump in its race for growth. In 1871, prices for refined oil products had plunged so low that the industry fell into a panic. Even Standard lost money at an alarming rate. The situation was so dire that for the only time in his life Rockefeller began to lose his nerve. He sold some of his stock and warned his wife that they would have to live on money from his other investments. Later in life he recalled, "All the fortune I have made has not served to compensate for the anxiety of that period."

There were no laws to regulate the quality of kerosene. If a kerosene contained too many impurities, it would explode inside the lamp when it was lit. In the mid-1870s, 5,000 to 6,000 people were killed each year in accidents caused by faulty kerosene. By naming his company Standard Oil, Rockefeller attracted anxious customers with an image of safety and reliability.

Standard's War on the Oil Industry

Standard recovered, but the experience shook Rockefeller. He saw the fierce competition among oil companies as wasteful and ruinous. In his view, the oil industry would be far more profitable and efficient if companies cooperated rather than competed. During the 1870s, he set out to gain this cooperation by reorganizing the entire industry under his control.

At first, this simply meant buying out competitors and, in some cases, inviting them to join his company. With his reputation for honesty and the fair prices he offered, Rockefeller had no trouble making many deals. In one four-week flurry in 1872 that historians call "The Cleveland Massacre," Standard Oil bought 23 companies, including 18 refineries.

When methods of persuasion and enticement failed, Rockefeller resorted to less noble tactics. Anyone who opposed Standard Oil could expect to get what Rockefeller called "a good sweating." A favorite Standard tactic was to concentrate on the strongest competitor. Standard would cut its prices in that competitor's market to below cost to drive the competitor out of business. Rockefeller also found ways to entice influential people to his side. He offered large amounts of stock to Cleveland bankers, who then found it to their advantage to give favorable loans to Standard Oil or refuse money to Standard's competitors.

Since Standard provided a large volume of business that the railroads desperately needed, it was able to negotiate better freight rates than its rivals. Standard added insult to injury by conniving with the railroads to create a system of "drawbacks." Drawbacks were fees that the railroads charged to Standard's competitors and then shared with Standard. By far the most controversial of Standard's practices, the drawbacks widened the company's already significant financial advantage by forcing competitors to finance Standard Oil unknowingly.

Throughout the 1870s, however, Standard operated so secretly that no one outside of Rockefeller's offices knew what the company owned or what tactics it employed. It bought companies anonymously or under the names of people not known to be associated with Standard. As a result, Standard came close to achieving Rockefeller's goal of complete control over the industry before the country even knew it was happening. By the end of the 1870s, Standard controlled more than 90 percent of the nation's oil refining.

In their quest for "efficiency," Rockefeller and his associates created an extraordinary system of corporate espionage. They tracked their competitors' moves so closely that they knew where each barrel of oil processed by each company went. They planted false stories in the press to influence the market. They bribed officials to the extent that one writer joked that Rockefeller had done everything to the Pennsylvania legislature but refine it.

When Standard met with an occasional setback, it kept applying pressure and making deals until it won. In 1879, some of Standard's competitors joined together in a company called Tidewater and devised the innovation of a long-distance pipeline from the oil fields of Pennsylvania to the refineries. This pipeline would drastically reduce transportation costs. Realizing Standard Oil might lose its advantage, Rockefeller tried everything short of violence to block the pipeline. He flooded manufacturers with fake equipment orders that kept them too busy to supply pipeline materials to Tidewater. He bought

Because companies were legally forbidden to own property in other states, businesses that became part of Standard Oil had to hide all evidence of the relationship. They continued using their original names, kept secret accounts, and corresponded with Standard in code. This secrecy was so effective that Standard executives worried every time the owner of a newly acquired refinery died; they were afraid his heirs might mistakenly try to claim the business for themselves, unaware that it belonged to Standard.

land in the pipeline's path, and he even called on his government contacts to oppose the plan.

Yet the Tidewater pipeline went through. Unable to beat his rivals, Rockefeller joined them and then pushed them aside. Not only did he construct his own pipelines, but eventually he also bought control of the Tidewater pipeline.

The Righteous Villain

monopoly: exclusive control over a product or service by one company, discouraging competition from other companies

In the late 1870s, Standard's colossal size finally attracted the attention of reporters and government officials. When the public learned of the oil empire Standard had established and the tactics it used, people were outraged. In 1879, a Pennsylvania grand jury indicted the company on charges of conspiracy to create a monopoly. Rockefeller managed to avoid

This New York Herald *political cartoon illustrates the widespread resentment of wealthy industrialists like Rockefeller in the late nineteenth century. The rich man is portrayed as a fat bag of money labeled "monopoly" who tries to keep his fortune—including the Standard Oil Trust—from the poor worker.*

going to trial, but he was branded for the rest of his life as a ruthless, money-grubbing villain.

In some ways, Rockefeller was the miser that his critics claimed him to be. He added up every bill to the penny to make sure he was not overcharged. Once he invited a Cleveland businessman and his wife to stay at his estate for the summer. At the end of their stay, the couple was surprised to receive a large bill from Rockefeller for their room and board.

Henry Morrison Flagler (1830-1913) was Rockefeller's partner in the founding of Standard Oil and remained a director of the company until 1911. He was also one of Rockefeller's closest friends, believing that "a friendship founded on business is better than a business founded on friendship."

Rockefeller was also seen as a remote figure whose only close friend was Henry Flagler. Many who met him were haunted by his cold "gun-barrel blue" eyes. The press became obsessed with Rockefeller's faults and exaggerated them so much that many Americans considered him the devil in a business suit.

Rockefeller, however, never wasted time answering his critics. In his own mind, he was the very model of an honorable, upright, moral human being. In fact, he was a devoted husband who involved his wife in his business decisions and encouraged her to give millions of dollars to social causes such as the advancement of black Americans. Contrary to his reputation as a workaholic businessman, he preferred to spend evenings and weekends with his son and his three daughters. He genuinely liked children, and would entertain them with such antics as balancing a plate on his nose.

Rockefeller never flaunted his wealth or drew attention to himself. He lived quietly near Cleveland for most of his life before moving to a secluded estate on the Hudson River in New York. A man of simple tastes, he wore the same suits until they were so shiny they had to be thrown out. Although Rockefeller contributed millions of dollars to establish the University of Chicago, he insisted that none of the buildings on campus be named in his honor.

Uncanny Foresight

Despite the storm of criticism, Standard Oil continued to prosper in the 1880s. In 1882, Rockefeller reorganized 40 of the companies he controlled into the great Standard Oil Trust. This new structure allowed Standard to acquire other oil companies without violating monopoly laws. Standard's individual stockholders would buy stock in a company and give their shares to Standard to manage for them "in trust." Since Standard Oil did not own the companies directly but only "cooperated" with them, the trust was technically legal. Often, companies held by Standard would, in turn, control other companies, which themselves controlled other companies. In this way, throughout the rest of the century, Standard Oil completely dominated not only the U.S. market but also the world kerosene market.

Rockefeller was not a particularly knowledgeable investor. His private investments were such a scattershot collection of hunches—some good, many bad—that he eventually brought in others to handle his personal finances. But he often had uncanny foresight when it came to large business deals. In the mid-1880s, prospectors discovered oil in northwestern Ohio near the town of Lima. The crude oil, however, reeked of sulfur. Try as they might, refiners could not eliminate that rotten egg smell from the oil, and so they could not interest consumers in products made from it.

Whereas all other refiners wrote off the Lima "skunk juice" as worthless, Rockefeller saw a golden opportunity to get into the production side of the oil business. When his associates objected to his plans to acquire land near Lima, Rockefeller offered to use his own money. Persuaded by Rockefeller's boldness, Standard spent millions of dollars buying up the Lima oil fields, then hired renowned chemist Herman Frasch to attack the problem of the odor. Within a few years, Frasch found a way to remove the smell. Almost overnight, the 40 million barrels of Lima oil that Standard had stored away rose in value from $4.5 million to more than $40 million.

The Lima, Ohio, oil field was the first new field that had been discovered in almost 10 years. Some oil industry analysts believe that if Rockefeller hadn't been so determined to make Lima oil usable, the United States may have suffered a critical oil shortage until the Texas boom 15 years later.

The "Living Mummy"

Standard Oil continued to grow beyond all of Rockefeller's expectations. By the turn of the century, Standard not only dominated oil refining but also had become the world's largest oil producer. Rockefeller was on his way to being the richest private citizen in the world.

In 1907, Standard was more than 20 times the size of Pure Oil, its nearest competitor.

During the later years of his life, Rockefeller probably gave away more money to worthy causes than any businessman before him. Yet the public never forgave him for the methods he used to rise to power. The stress of building and maintaining an enormous operation in the face of such hatred took its toll on Rockefeller. In 1901, he developed a condition called alopecia that caused all of his hair, even his eyebrows and eyelashes, to fall out. Ridiculed in the press as a "living mummy," Rockefeller kept to himself for most of his later life. He died in 1937, six weeks before his 98th birthday.

Legacy

Rockefeller was pleased, but not terribly impressed, by the enormous empire he had created. "Wealth isn't a distinction," he once said. "If I have no other achievement to my credit than the accumulation of wealth, then I have made a poor success of my life."

Rockefeller firmly believed that by seizing control of the chaotic oil industry, he had created a prosperous business that greatly benefited people around the world. While Standard's production efficiency made many refined oil products more available and affordable for the average consumer, not everyone shared his opinion. Many Americans, alarmed by the dramatic industrialization of the late nineteenth century, saw big businesses such as Standard Oil as impersonal monsters that destroyed lives and careers in the name of progress.

Standard Oil became so powerful that the U.S. government finally acted to tame the giant. In 1906, the government filed suit against Standard, claiming that its tactics violated the Sherman Anti-Trust Act of 1890. On May 15, 1911, the U.S. Supreme Court agreed. It ordered the Standard Oil Trust dissolved into 38 separate companies that were to operate independently of each other.

In the coming years, these companies continued to prosper, and they grew into some of the most

Rockefeller (second from left) on his way to court to testify on behalf of Standard Oil in a 1908 antitrust case

widely recognized businesses in the nation. Standard Oil of New Jersey became Exxon, Standard of New York became Mobil, Standard of California became Chevron, Standard of Indiana became Amoco—and the list goes on. If the court had not ordered the disbandment of the Standard Oil Trust, Rockefeller's creation would be far and away the world's largest corporation.

Ironically, Rockefeller's legacy is primarily the "accumulation of wealth" that he downplayed. Although he developed the model for today's large-scale, multinational corporations, he created no new products or innovations. He simply outcompeted all rivals to forge the most powerful business empire in history.

2

Marcus Samuel

Shell Takes on the Giant

Had the oil experts known what Marcus Samuel was planning in 1890, they would have been aghast at what appeared to be a death wish. The relatively obscure, small-time trader was preparing to challenge massive Standard Oil, a move that seemed equivalent to a mouse taking on a lion. Standard controlled over 90 percent of the world's refined oil, and it could be brutal to those who posed any threat to its profits.

Furthermore, Samuel's plan was daring beyond belief. Rather than nibbling at a small part of Standard's kerosene market, which was all that Standard's remaining competitors could do, Samuel decided to declare all-out war. He would try to push Standard aside and grab control of the oil market throughout the continent of Asia.

Sir Marcus Samuel (1853-1927) created the Shell Transport and Trading Company with the belief that "The mere production of oil is almost its least value and its least interesting state. Markets have to be found."

Objects decorated with shells were fashionable in Victorian homes. At the peak of his career, Marcus Samuel Sr. employed designers, several shell cleaners, and 40 young women to manufacture products like this box.

A Business From Seashells

Marcus Samuel had risen a long way from his beginnings in the slums of London's East End. He entered the world on November 5, 1853, the 10th of 11 children born to Marcus and Abigail Samuel. The Samuels were devout Jews, which meant they faced a good deal of discrimination.

But the Samuel family did not live in the slums because poverty forced them to. On the nearby East London docks, the elder Marcus Samuel was able to build a modest fortune buying and reselling seashells and other trinkets that sailors picked up abroad. He used the shells to decorate jewelry boxes, picture frames, and ornaments, and these items soon became his most popular products.

By the time Marcus Jr. was four years old, profits from the shell trade enabled the Samuels to move to a more comfortable neighborhood. The business gradually expanded into a full-scale trading house, with many clients and contacts in Asia.

The younger Marcus Samuel was a shy, soft-spoken boy, somewhat overwhelmed by his many older brothers and sisters. Growing up near the docks had given him a love of ships and the sea that would last throughout his life. Marcus began working in the family business at age 16. But his oldest brother, Joseph, who took charge of the company after their father's death in 1870, had difficulty treating his baby brother as a grown-up. After two years of laboring under him, Marcus could stand it no longer. At age 19, he sailed off to Asia to learn about the family business for himself.

Breaking into the Far East

In 1873, Samuel arrived in an area of India that had just been hit by a severe famine. Overwhelmed by the need he saw, he overcame his natural shyness and went to work to solve the problem. Aided by his father's good reputation among Asian merchants and traders, Samuel was able to charter a ship that brought in 50,000 tons of rice from Siam (now Thailand), feeding the hungry while earning him a tidy profit. That episode taught Samuel the basic principle on which he would run his business: seek the nearest source of supply to meet local demand.

Samuel was called home in late 1874 and arrived just in time to see his mother before she died. Again, he tried to work with Joseph in the family business. He and his younger brother Sam quickly grew frustrated by the unimaginative way the company was run and their lack of input into decisions. Marcus made two more voyages to the Far East in the late 1870s, establishing a network of business contacts throughout Asia. Finally, he and Sam put together two trading companies—M. Samuel and Company in London and Samuel Samuel and Company in Japan. By the time Marcus Samuel married Fanny Benjamin, from another family of trade merchants, in January 1881, he and Sam were well established in their business.

Marcus and Sam

The two brothers kept their operating costs to the bare bones. They worked together in a cramped office in a cluttered warehouse in East London, equipped with only a table, two chairs, and a map of the world. Despite the intricate deals they struck with a wide variety of customers from many countries, their office staff consisted of only two clerks to help with the paperwork.

"The two brothers would always go to the window, their backs to the room, huddled together close, their arms round each other's shoulders, heads bent, talking in low voices, until suddenly they would burst apart in yet another dispute, Mr. Sam with loud and furious cries, Mr. Marcus speaking softly, but both calling each other fool, idiot, imbecile, until suddenly, for no apparent reason, they were in agreement again."
—a Samuel employee

Marcus was the ideas man of the two; Sam was better at figuring out how to carry out a plan. The brothers seldom agreed on any new course of action. In contrast to the typical corporate image of sterile, number-crunching boardroom meetings, the Samuels' business conferences were desk-rattling free-for-alls that left eavesdroppers aghast. After ridiculing each other and all but coming to blows, they would suddenly reach an agreement. The echoes of enraged name-calling barely faded before observers would see the brothers with arms around each other's shoulders as they worked out the details of their new plan.

Although Marcus and Sam achieved great wealth before either reached the age of 30, they kept looking for grander and more profitable business deals. The boldest of them all was a plan to challenge Standard Oil for the Asian kerosene market.

Following his brother Marcus's lead, Samuel Samuel (1855-1934) balanced his business interests with an involvement in politics. Eventually, he became a member of Parliament.

RUSSIAN OIL

In the late 1880s, the only major source of oil for kerosene outside the United States was in the Baku oil fields of southwestern Russia, near the Caspian Sea. This oil was produced by the Nobels—a Swedish family of entrepreneurs, one of whom created the famous Nobel prizes—in partnership with the French banking family, the Rothschilds.

The Samuel brothers dabbled in buying and selling some of this oil in Asian markets. Their research suggested there was a great demand for kerosene in Asia, including some areas in which it was not currently available. When Fred Lane, a shipper with whom the brothers had often done business, approached the Samuels to see if they wanted to get into oil shipping on a larger scale, Marcus listened. Unfortunately, one huge obstacle loomed over the project: the major provider of kerosene in the world was Standard Oil. No one in their right mind wanted to tangle with that invincible giant.

Marcus Samuel visited the Russian oil fields in 1890. There he saw a primitive oil tanker—a ship with a large storage compartment for carrying oil. Samuel knew that Standard shipped its oil products in tin cans. Samuel realized at once that he could save a tremendous amount of money by shipping kerosene in bulk rather than packaging it first. The problem was that the most practical shipping route from the Russian oil fields to Asia was through the Suez Canal, built by the British on Egyptian territory. The British authorities in charge of the Suez feared that a large ship loaded with flammable kerosene maneuvering in a narrow canal was an invitation to disaster. They had recently denied Standard's requests for permission to send its tankers by that route. Samuel believed his only chance of success was to design a new, safe tanker that could win approval to pass through the canal.

A map of Samuel's proposed tanker route. Passing through the Suez Canal reduced the journey between England and Asia by 4,000 miles.

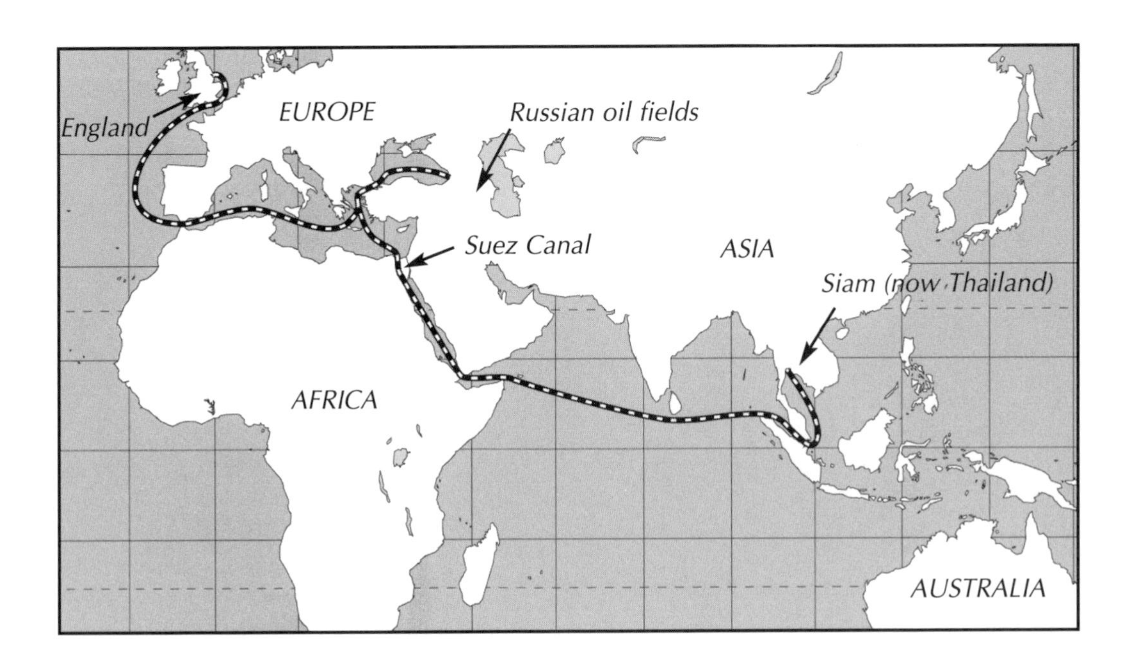

Going For Broke

Samuel also came to the daring conclusion that the only way to beat the giant Standard was to wage all-out economic war against it in all markets. He saw that if he tried the safer strategy of competing against Standard on a small scale in a few markets, he could not win. Standard would drop its price below cost in those markets and make up for lost revenue by raising prices in markets in which it had no competition. Samuel would be able to match Standard's price only by selling kerosene for less than the cost of production, but he would quickly go broke. If, however, Samuel challenged Standard in all Asian markets, he, too, could raise prices in one market to offset a price war in another market.

Secrecy was the key to his success. Samuel had little chance of establishing new markets unless he could grab them before Standard knew what was happening. For several years, Samuel prepared for the bold strike against the most feared business competitor in the world. With backing from wealthy investors and partners, he secretly sent two of his nephews on a tour of southeast Asia to buy land at key distribution points. The storage tanks he installed at these locations would allow him to get kerosene to nearby markets quickly and cheaply.

Samuel determined what safety features the Suez Canal directors would require in an oil tanker before they would allow it to pass through the canal. Then he secretly ordered a fleet of custom-made tankers to satisfy those requirements. These tankers, the largest the world had ever seen, included expansion tanks so that the kerosene, when heated during the course of a long journey through warm climates, would not build up to dangerous pressures that might cause it to explode. Samuel also saw that he could increase his profits if he could ship products back to Russia on the tankers' return trips rather than sailing empty. He developed a means of steam-cleaning the cargo holds so thoroughly that he could even ship foods such as rice, tapioca, and sugar in the same compartments that had carried kerosene. Samuel's confidence in his new ships was so strong that in 1891, after long negotiations, he signed a nine-year contract with the Rothschilds to sell the oil products from the Russian fields. The next year, he finally won passage through the Suez Canal.

By the end of the year 1895, 69 tanker passages had been made through the Suez Canal. All but 4 of the ships making the journey were owned or chartered by the Samuel brothers.

In July 1892, Samuel was ready to launch the first of his new oil tankers. He named the ship the *Murex*, after a type of seashell—in honor of his father's business that had started the family fortune. In the following year, Samuel launched 10 more tankers, all named after seashells.

Standard's Counterattack

Samuel succeeded in catching Standard completely off guard. The giant corporation tried desperately to strangle its new competitor. It fought hard in court and in the back halls of the British government to prohibit Samuel's ships from passing through the Suez Canal. Some of its supporters tried to play on racism, asking officials to stand firm against the "Hebrew influence" that sought control of the canal.

Samuel's first tanker, the Murex, *made 25 passages through the Suez Canal by the end of the nineteenth century.*

Fortunately, Samuel was heavily involved in local British government and had recently been elected a London alderman. His honorable reputation and his contacts in government helped derail all attempts to ban his ships from passage through the canal. Samuel's head start in designing safe oil tankers gave him a tremendous advantage for more than a decade. By 1902, approximately 90 percent of the oil passing through the Suez Canal belonged to Samuel.

Having failed to block Samuel's passage through the canal, Standard then launched a furious attack against him in the Far East markets. It slashed its prices drastically throughout Asia. The price war devastated the industry. Hundreds of small oil traders and producers lost so much money that they went out of business. But because of the shorter

An Egyptian kerosene salesman fills his cart from a storage tank. Samuel's network of tankers and supply depots helped his company compete with Standard Oil throughout the world.

distance his tankers had to travel compared to Standard's ships, which carried oil all the way from Pennsylvania, and the network of supply depots he had set up, Samuel's costs were much lower than Standard's. He could earn a profit selling at prices that were hurting Standard.

Samuel's project, however, ran into an unexpected hitch in several Far Eastern cities. The plan had been to keep costs down by shipping in bulk and having customers bring in their own containers that Samuel's company would fill from its tanks. But many customers refused to buy kerosene from Samuel and instead bought from Standard, even at higher prices. Samuel was baffled until he found out that many customers considered Standard's tins to be even more valuable than the kerosene they contained. Consumers used these tins for everything from tea strainers to roofing material.

Samuel quickly loaded a ship with tinplate and sent it to his business partners in the Far East with instructions to start making kerosene containers. His partners complained that they didn't know anything about manufacturing containers, but Samuel offered no sympathy. The containers had to be made, so the company officials had better figure out how to make them! They did, and sales of oil increased dramatically when the containers became available. Once again, Samuel's policy of using the nearest source of supply paid off. By the time Standard's blue kerosene tins arrived in Asia, they were chipped and dented from being shipped halfway around the world. But because Samuel's red tins were made on site, they were bright, shiny, and much more attractive to customers.

Samuel at the Top

In the mid-1890s, Marcus Samuel was riding high. Not only was he winning his daring challenge to Standard, but he was also making a huge profit supplying weapons to Japan for its war with China. The child of the London slum now lived comfortably on a 500-acre country estate. His children were enrolled at England's finest schools. Whereas his father had been banned from public office because he was Jewish, Marcus was fast becoming a force in London politics. In 1894, he added Sheriff of London to his growing list of prestigious government positions. Four years later, he was knighted by Queen Victoria

In 1893, Marcus Samuel became desperately ill with what physicians diagnosed as cancer. In fact, his doctors warned him that he was not likely to survive six months. But Samuel made a remarkable and complete recovery.

This photograph of Marcus Samuel's luxurious house, The Mote, was originally featured in The Jewish World *newspaper in 1913.*

after one of his tugboats set free a British warship that had run aground at the entrance to the Suez Canal.

Samuel felt so flush that he turned down offers from Standard that would have guaranteed him a life of ease and luxury. Having failed to squeeze out Samuel, Standard reached into its vaults and tried to buy him out. After Samuel refused a prestigious position on Standard's board of directors, the desperate company offered him $40 million. More out of patriotism than anything else, Samuel rejected the offer. He hated the idea of the entire world being at the mercy of an American company for its oil products. He thought Great Britain needed his company to guarantee a supply of oil.

Shell Transportation & Trading Company

While enjoying success, Marcus began to worry about the danger of relying on a single Russian source for oil. If that should dry up or if the price should rise dramatically, the Samuels would have no income to pay off their huge investment in the oil business. Sure enough, the Rothschilds, primary owners of the Russian oil fields, demanded more and more for their oil.

Samuel then set out to seek a new and closer source of kerosene for Asia. In 1895, he bought oil rights to a large area of uninhabited, overgrown jungle on the eastern coast of Borneo. He then sent his inexperienced nephew, Mark Abrahams, to supervise the search for oil. As with the tin manufacturing

A derrick at Balik Papan in Borneo. Visible in the background is the thick jungle that made it so difficult for Samuel's workers to drill for oil there.

venture, Marcus and Sam Samuel offered much criticism but little help or direction to Abrahams. Working in a blistering, steamy jungle a thousand miles from the nearest supply port, Abrahams struggled valiantly. Many workers died in the quest, but in 1897 the team finally struck oil. The oil proved to be thick and heavy—good only for fuel products and not kerosene. At this time, there was little demand for oil as a fuel. But Samuel had seen oil used to power ships near the Russian oil fields where wood and coal were scarce. He correctly predicted the "tremendous role which petroleum can play in its most rational form, that of fuel," and went into fuel oil production.

That October, the Samuel brothers gathered all of their tankers, storage, and production facilities into the Shell Transport and Trading Company, named in honor of their father's business origin.

Shell Slips Away

The Samuel brothers found that maintaining a huge company was far more difficult than building it. Although Marcus was a man of great vision and ideas, neither he nor his brother had much talent for organization or administration. Their haphazard style of operation left them helpless to deal with a rash of problems that began to arise. In 1900, the Boxer Rebellion in China trashed Shell's facilities. Shipping rates fell, draining income from the Shell fleet. Fatal explosions and inefficient design plagued production and refining in Borneo. A supplier failed to fulfill a contract for a huge supply of Texas oil.

The British navy rejected the use of oil as a fuel for ships in favor of coal.

Upon being elected Lord Mayor of London in 1902, Marcus had little time to spend on these problems. After a world slump in oil prices in 1903 rocked Shell to the core, the company deteriorated rapidly. Disgusted by the turn of events, Fred Lane resigned from the company, telling Samuel, "You are, and have always been, too much occupied to be at the head of such a business." By 1907, Shell was forced to accept a merger with the Royal Dutch Oil Company. Samuel could only stand by and watch as control of his company slipped away to Royal Dutch's more aggressive leader, Henri Deterding.

merger: an agreement that combines two or more corporations into one

Marcus Samuel was only the third Jew to be elected Lord Mayor of London, the highest local government post in England.

Sir Marcus Samuel continued as a public servant and active shareholder in Royal Dutch/Shell into the 1920s. He lobbied hard for the British navy to adopt liquid fuel for its ships, which it officially did in 1921. That same year, Samuel was made a nobleman, and in 1925 was named Viscount Bearsted for his many accomplishments. He died in 1927 at age 73, only 14 hours after his wife's death.

Legacy

The Royal Dutch/Shell Group has the distinction of being the company that reintroduced competition into the international oil business. Standard Oil's grip on the worldwide industry was so tight and its tactics so aggressive in the late nineteenth century that the company appeared invincible. The world was faced with the frightening prospect that a single company would control the most important energy resource in the world. Marcus Samuel, with brother Sam at his side, was the first to mount a full-scale challenge against Standard and survive.

Under the direction of Henri Deterding and those who followed him, Royal Dutch/Shell not only opened up markets and oil fields in southeast Asia. It also pioneered oil exploration in South America with its 1922 discovery of a major oil field near Lake Maracaibo in Venezuela. Buoyed by this venture, Royal Dutch/Shell has competed all over the world, selling its products under the familiar yellow shell logo. Yet it has kept its corporate workings behind the scenes. This has had an unusual result: few of its customers are aware of Royal Dutch/Shell's origins

and mixed nationality. Over the course of the century, most Americans filling up at a Shell Oil station in the United States might have believed they were buying gasoline from an American company. At the same time, most British customers have assumed that the company is British, while customers in the Netherlands have viewed it as Dutch.

Although the company has international origins, Shell gasoline stations are familiar sights throughout much of the United States.

Royal Dutch/Shell has grown into a corporate giant that, along with British Petroleum, continues to provide the stiffest international challenge to the American oil companies. Like the world's other giant oil companies, Royal Dutch/Shell has become, in the words of chairman David Barran in 1967, "so international that nothing can happen in any part of the world without it affecting our interests."

In 1999, the Royal Dutch/Shell Group ranked as the world's 11th largest corporation, with revenues of almost $94 billion.

3

Pattillo Higgins & Andrew Mellon

Gulf Oil: The Spindletop Boom and Bust

While oil brought riches to Pennsylvania throughout the latter half of the nineteenth century, Texans could only look on in envy. Few industry experts dreamed that Texas might hold an even greater pool of oil than Pennsylvania.

Yet, had they studied their history, they would have known that in 1543 the ragged remains of Hernando DeSoto's Spanish expedition reached Texas and discovered a sticky black substance seeping up from the ground. To the explorers' pleasant surprise, this oily material was ideal for patching their leaky boats. Further research would have shown that, for centuries, the Native American tribes living

Both Pattillo Higgins (top), the rowdy son of a blacksmith, and Andrew W. Mellon (1855-1937), a wealthy banker, played important roles in the complex history of the Gulf Oil Corporation.

in Texas had rubbed the same thick liquid on their bodies to cure aches and pains.

A few enterprising Texans had made feeble attempts to drill for oil on their property. But by the mid-1890s, the yearly oil production of the entire state of Texas was 60 barrels—less than a single small Pennsylvania well produced in a month.

Pattillo Higgins and the Leaking Gas

It all changed thanks to Pattillo Higgins, a rough-and-tumble blacksmith's son born in 1863 in Beaumont, Texas. Higgins's missing arm, said to have been lost in a gunfight, served as a permanent reminder of his wild, unpredictable nature. Unable to settle into a career, he bounced around from job to job as a railroad worker, logger, mechanic, and cabinet maker. Yet Higgins was also a man of ideas who read avidly and taught Sunday school.

In the late 1880s, Higgins tried yet another job, this time at a brick-making plant. While on a business trip to Ohio for his employer, Higgins visited a factory that used oil for fuel. He stared in fascination. Oil burned with an even heat that was easy to regulate, and it was obviously far superior to the wood or coal fires Higgins was used to. But back home, people only thought of oil as worthless, skunky-smelling stuff that sometimes seeped up through the ground.

After visiting several eastern oil fields, Higgins wondered about the low hill, a half mile in diameter, that rose about 12 feet above the swampy flatland outside Beaumont. Most locals simply referred to it as Big Hill, but it would later come to be called Spindletop. Higgins remembered taking his Baptist Sunday school class out to a foul-smelling spring near Spindletop, where sulfurous gas bubbled up through the water. He had amused the children by touching a flame to the gas bubbles and watching them ignite. This gas seepage, he thought, might indicate the presence of oil. Higgins started reading everything he could find about the subject of geology, and his research convinced him that the unique formation of the low hill was another sign of oil. He bought 1,770 acres of land near Spindletop—land that would be worth millions if his hunch proved to be correct.

Higgins spoke so persuasively about the oil potential of Spindletop that some Beaumont residents joined him to form the Gladys City Oil, Gas, and Manufacturing Company in 1892. Higgins named the company after Gladys Bingham, a girl from that Sunday school class he had entertained.

While Higgins was raising money for his oil well, the small town of Corsicana, about 130 miles northwest of Beaumont, was drilling three new water wells. The first well reached slightly more than 1,000 feet when it began filling with oil. Dismayed, the drillers drained the contaminating oil into a large pit near the well and continued drilling for water. A few enterprising townsfolk, however, thought the oil might have commercial possibilities. They were right, and by 1897 Corsicana had become the first Texas oil field, producing over 65,000 barrels a year.

Nothing But Quicksand

The Gladys Company hired a driller in 1893 and waited eagerly for the results. Unfortunately, by 400 feet the driller had hit nothing but a large pocket of quicksand. The standard drilling method of pounding a sharp bit into the earth proved virtually useless in the heavy, shifting sand.

Higgins and his company could not scrape together the money for a second try until 1895. This effort, too, bogged down in the wet sand. A third attempt in 1896 proved no more successful.

Stubbornly, Higgins held to his conviction that oil lay beneath Spindletop. The success of the deeper wells in Corsicana to the north convinced him that his only failure was in not digging deep enough. But by this time, the townsfolk had lost faith in him. Most concluded he was crazy. Desperate to keep the project going, Higgins hired geologists to study Spindletop and offer their opinion of its oil potential. The result was the opposite of what Higgins hoped. None of them saw any reason to believe there was oil at Spindletop. One geologist warned people "not to fritter away their dollars in the vain outlook for oil in the Beaumont area." After that report, no one would invest in the grand oil scheme. Higgins was forced to sell most of his land to pay his debts, and he became the butt of jokes in town.

Anthony Lucas brought his knowledge of engineering and his experience in salt drilling to the search for oil at Spindletop.

Anthony Lucas

Having run out of ideas and down to his last 33 acres, Higgins tried a blind appeal to strangers. He bought an advertisement in an eastern business journal in which he told of this mound leaking sulfurous gas and asked for a partner to help him drill for the oil that must be there.

Only one person, Anthony Lucas, responded to the ad. Lucas, an engineer and former naval officer, could sympathize with Higgins's dream. Born in Austria in 1855, he had come to the United States to prospect for gold in 1879. But after several months of finding nothing, he gave up and turned to the slightly less risky profession of drilling salt mines in Louisiana.

Two things about Higgins's description of Spindletop intrigued Lucas. First, the smell indicated that the land probably contained a good amount of sulfur, which Lucas was interested in mining. Second, the shape of the hill suggested the presence of a salt dome. From his experience as a salt driller, Lucas had concluded that oil often lay under salt domes.

Finding that Higgins had virtually no money left to fund the project, Lucas arranged a deal. He bought the Gladys Company's land for $33,150 and agreed to pay Higgins 10 percent of his profits from the oil. Lucas began work in 1899, hauling in drilling equipment from Louisiana. So confident was he of success, despite Higgins's six years of failure, that he brought his wife to live with him in a shack near the hill. He expected that the inconvenience of living with egg crates and apple boxes as furniture would last only a few weeks.

Six months later, however, the Lucases were still in the shack. Lucas's drill was about where Higgins's had been six years earlier—bogged down in the wet sand. He had found precious little oil or sulfur. The piping in the hole had collapsed under gas pressure. With his money running so low he could barely afford food, Lucas looked around for help—this time from the people who really knew the oil business, Standard Oil. But Standard's investigator did not buy Lucas's theory of oil in salt domes and reported to his supervisors that he saw "no indication whatever to warrant the expectation of an oil field on the prairies of southeastern Texas."

Over 200 million years ago, hot, dry conditions created large deposits of salt that are now buried as deep as 45,000 feet underground. In places where the salt is lighter than the soil surrounding it, it has pushed its way upward, creating a dome on the earth's surface. As the salt rises, it tilts layers of sedimentary rock, shifting oil and gas deposits and trapping them in large pools. Thus, major oil discoveries are often made near **salt dome** formations.

Galey and Guffey

Lucas decided to try one last expert, Dr. William Battle Phillips, professor of geology at the University of Texas. Supporting Lucas's theory about the connection between salt domes and oil, Phillips agreed that Spindletop had oil potential. More importantly, he gave Lucas a letter of introduction to John Galey and James Guffey, one of the most respected drilling teams in the country.

Rotary drills had been used to drill water wells since the 1870s but did not arrive in Texas until 1895. There, oil drillers found them ideal for drilling through quicksands and soft rock. Unable to compete with efficient rotary drills that could dig a 1,000-foot well in just 36 hours, cable-tool drills soon vanished from the oil industry altogether.

Galey, whose instinct for finding oil was legendary, inspected Spindletop and predicted it held the largest oil field in North America. He and Guffey agreed to take over all expenses of the project from Lucas in exchange for seven-eighths of the profits. Lacking the money to finance the project themselves, Galey and Guffey borrowed $300,000 from the Pittsburgh bank of T. Mellon and Sons.

Drilling began on October 27, 1900. To combat Spindletop's quicksand, Galey and Guffey brought in a rotary drill, which twisted the drill bit deep into the ground like a screw instead of pounding it in like a nail as conventional cable-tool drills did. A recent innovation in oil drilling, rotary drills had already proved successful at reaching oil in soft ground. The crew also found a clever way to prevent the cave-ins and water and gas leaks that slowed work down: they drove cattle into a shallow lake and used the mud they stirred up to seal the sides of the drilling hole.

Working 18-hour shifts, the crew was exhausted by the time they reached 880 feet on Christmas Eve. As yet, they had uncovered only a small amount of oil, which would be difficult to refine into a useful product because it contained fine sand.

According to one source, Galey considered abandoning the effort to drill in another spot, but he was stopped short by Mrs. Lucas. "Mr. Galey, the contract calls for drilling this well to twelve hundred feet," she reminded him. "We need to know what there is down that far."

GUSHER!

After a break for the holidays, the crew drilled down past 1,000 feet. Shortly after 10 A.M on January 10, 1901, they were installing a fresh drill bit when mud started spouting up from the ground. Seconds later, six tons of drill pipe shot out of the hole and smashed through the top of the derrick. Stunned, the crew surveyed the site, which was covered with debris and a layer of mud half a foot deep. As they were wondering how long it would take to repair the damage, more mud exploded from the hole. Then rocks burst hundreds of feet in the air, propelled by an enormous jet of oil. Over 75,000 barrels of oil roared out of the ground in that first day. Not only had no one ever seen anything like it, no one had even imagined such a gigantic fountain of oil.

The discovery brought chaos to Beaumont. Within a week of the strike at Spindletop, more than 1,000 people swarmed into town to gawk at the "freak" gusher, and at least 40,000 more had followed by the end of the year to try their luck in the oil business. Land bought for $10 an acre two years earlier now sold for $900,000 as speculators fought

When oil began to show in the Spindletop well, the crew's lead driller excitedly estimated that the oil pool might yield 50 barrels a day, twice as much as the wells he knew of in nearby Corsicana. In fact, the Spindletop gusher produced 3,000 times more than any Corsicana well ever had.

lease: a contract granting the use of property for a specified time in exchange for payment. The land on which a person or company drills for oil is often leased from private owners, who receive fees or royalties.

to lease land near Spindletop. One tiny plot selling for $8 in the morning changed hands dozens of times until the last buyer of the day paid $35,000 for it. By 1902, 100 companies operated 440 gushers crammed together on the hill, and Spindletop was producing more oil than the rest of the world combined.

Among the new oil companies was the J.M. Guffey Petroleum Company, which Guffey formed after buying out Galey's and Lucas's shares in the venture. The Mellon bank was again Guffey's major

The largest well ever struck before Spindletop, in the Baku fields of Russia, had gushed for 11 hours before subsiding. Ten days after Lucas struck oil in Texas, however, the fountain of oil was still going strong. To stop the oil from being wasted, Galey and Guffey installed a system of valves on top of the well to regulate the amount of oil that flowed from it. This procedure, called capping, became a common practice as more and more gushers were discovered in Texas.

The Gulf refinery towered over the cow pastures of Port Arthur, Texas.

investor. One of the company's assets was the Gulf refinery, 19 miles away in Port Arthur on the coast of the Gulf of Mexico. From there, the Spindletop oil could be shipped to markets around the world.

Guffey was a dramatic, blustery man who was good at selling and promoting but poor at organization. He was unable to deal with the economic cyclone swirling around him. His company wasted thousands of barrels of oil because of inadequate storage facilities and lost more money in the frequent fires that raged through the derricks. Meanwhile, the glut of oil had caused oil prices to plunge from several dollars per barrel to three cents per barrel—two cents cheaper than a cup of water.

The discovery of oil transformed Beaumont into what one historian calls "a seething, frantic concentration of fortune seekers, swindlers, gamblers, prostitutes, sellers and buyers of anything and everything." The area became known as "Swindletop" because of the endless frauds and speculation that occurred there.

Guffey made what seemed like a favorable deal with the Shell Transport and Trading Company. Shell contracted to buy a huge amount of the company's oil at a guaranteed price of 25 cents per barrel for 20 years. But then overproduction caused the great oil pool to lose pressure. Within two years of the first strike at Spindletop, its wells were running dry. Guffey suddenly had no great source of oil to sell. To meet the contract, he was forced to buy oil from other companies and then resell it to Shell. But since the shortage had caused prices to rise to 35 cents per barrel, he had to pay far more for the oil than he could charge to Shell for it.

The Rule of Capture

The law governing the early pursuit of oil was called the "rule of capture." This idea was first developed to settle disputes among game hunters over who had the right to kill migratory animals and birds. Under English common law, the owner of a piece of land also owned any wild creature on it, even if that creature's home was elsewhere. Applied to oil drilling in America, this rule meant that landowners had the right to whatever oil they could "capture" on their own land, even if the oil reservoir actually lay below another person's property. If several people owned land above a single pool of oil, the oil belonged to whoever got to it fastest and pumped it hardest.

The rule of capture played a major role in shaping the early oil industry. It transformed the search for oil into an exciting, highly competitive race in which speed and determination were everything. Because many wells could be drilled over an oil pool and each was just as likely to strike it rich, there was room for many people to enter the oil industry and make their fortunes.

The rule of capture, however, also caused a huge amount of waste and damage. In places like Spindletop, oil pools were drained so quickly that they produced far less oil than they might have. Oil reservoirs usually contain natural gas, which, when trapped underground, creates a pressure that forces oil to the surface. If the natural gas is released too rapidly, the oil stops flowing upward and remains underground, out of reach. Furthermore, oil wells were crowded so close together that fires started easily, spread instantly, and destroyed any oil already brought to the surface. As Anthony Lucas observed, Spindletop was like a cow that was "milked too hard. Moreover, she was not milked intelligently."

Overproduction at Spindletop caused millions of barrels of oil to be lost and, as a result, thousands of careers were destroyed. The rule of capture had created a boom-and-bust cycle that made the oil market very unstable. As everyone hurried to pump as much oil as possible from a newly discovered pool, the amount of oil on the market increased until it exceeded demand, and prices dropped. As soon as the pool was drained dry, oil was in short supply and thus incredibly valuable. Then a new oil field would be discovered and the cycle would begin again.

It was not until the 1920s that lawmakers and oilmen began to question the rule of capture and its effect on the economy. When the Great Depression struck in 1929, the government sought more control over industries to prevent economic chaos. Finally, in 1933, the Oil Code was established, placing a limit on the amount of oil produced by each state per month. The unbridled production encouraged by the rule of capture was over.

It was often said that the Spindletop wells were so close together that a person could walk from derrick to derrick without touching the ground. By the time this photograph was taken in 1903, the huge oil pool had been nearly exhausted by overproduction.

The Mellons Take Over

The reports coming from Texas worried the Mellons, who had pumped millions of dollars into Guffey's company. When it came to finances, the Mellons were tough customers. Thomas Mellon had started the bank in 1869 after making a fortune in Pittsburgh real estate. One of his less honorable tactics was making loans to struggling farmers, hoping to get farm property cheaply when the farmer failed to repay the loan. Thomas's son, Andrew, took over the bank when he was only 26 years old and built it into a financial power with the same strong-armed, heartless tactics his father had used.

Prior to Spindletop, the Mellons had little experience in the oil business. They had dabbled in oil only once before, in 1889, when they had bought an oil field outside Pittsburgh. Andrew Mellon had recruited his 19-year-old nephew, William Mellon, to run the business. William handled the job so well that the Mellons made a substantial profit when they sold the business to Standard Oil in 1895.

Calling the state of Guffey Petroleum in 1902 "just about as bad a situation as I had ever seen," William Mellon believed the company could only be saved by "good management, hard work, and crude oil."

Upset by Guffey's bungling of the oil interests in Texas, Andrew Mellon now sent William to investigate. William, who admitted, "It is very hard for me to be patient with incompetents," was appalled at the mess he found. He reported to his uncle that they would have to invest $15 million more into Guffey's company in order to make it profitable.

Andrew Mellon had no intention of wasting that much money on an oil company. As he had done with his Pennsylvania field, he offered it to Standard Oil. Standard, however, wanted nothing to do with Texas oil. Ever since Standard's size and business tactics became public knowledge, Texas government officials had been openly hostile to both John D. Rockefeller and his giant oil trust. Not only did they pass laws that Rockefeller thought targeted him unfairly, but they even tried to prosecute him as if he were a criminal. As a result, Rockefeller icily refused to do business in Texas.

Buckskin Joe and Texaco

When oil was discovered in Texas, the region had no facilities for storing, transporting, refining, or selling its new resource. Standard Oil's feud with Texas left the market wide open for small, local companies to develop the industry for themselves, and Joseph S. Cullinan took prime advantage of the opportunity. Starting his oil career as a crew boss in Pennsylvania, Cullinan earned the name "Buckskin Joe" because of an aggressive, hard-driving personality that reminded coworkers of the rough leather gloves used in the oil fields. He worked his way through the ranks at Standard Oil and then gave up a comfortable executive job to strike out on his own.

In 1897, Cullinan heard that the people of Corsicana, Texas, had struck an oil field but had no idea how to profit from it. When the town's mayor wrote to him for help, Cullinan moved to Corsicana and set to work managing the construction of storage tanks, pipelines, and the first refinery west of the Mississippi River. He traveled around the state, convincing businesses and railroads to switch from coal to oil for their fuel needs. Under Joseph Cullinan's direction, Corsicana became the oil capital of Texas.

But when oil began spurting from the ground at Spindletop, Cullinan saw an even greater chance to strike it rich with Texas oil. Within three months, he and a group of wealthy investors formed the Texas Fuel Company, which later became the Texas Company and was finally shortened to Texaco. The company spent lavishly for control of the most valuable property near Spindletop.

In 1902, the Texas oil industry would again turn to the experienced Cullinan for help. That September, an oil driller on a derrick at Spindletop carelessly flicked a lit cigar to the ground, and the entire oil-saturated area burst into flames. When asked to organize the fire-fighting effort, Cullinan agreed on one condition: he would enforce his orders at gunpoint. With dozens of businesses, including his own, on the brink of ruin, Cullinan worked tirelessly against the blaze. For a solid week, he stayed in the front lines of the effort, repeatedly risking his life as he directed the sand-dumping and steam-blasting to douse the flames. Although he ended up hospitalized with exhaustion, seared lungs, and temporary blindness, he accomplished his task and saved the oil field from total destruction.

But while Cullinan's rough, flamboyant, and sometimes overbearing style could be an advantage in the oil fields, it began to grate on investors who were used to calling the shots themselves. Gradually, these rich East-Coast investors maneuvered to take control of the Texas Company. Cullinan fought tenaciously, but in 1913 those with the money won out. Cullinan responded in the best tradition of the Texas gunfighter: "It was a good boarding-house brawl," he said, "and some furniture was broken but our side was whipped fair and I'll be looking for another job soon."

Although Joseph Cullinan once guaranteed that Texaco would always be based in Texas, the local flavor has long since faded from the company. Currently headquartered in White Plains, New York, Texaco has become a vast international enterprise. With annual revenues of more than $31 billion, it is the largest American competitor of the original Standard companies. But while its frontier spirit may have been lost, Texaco—like the oil industry itself—still remains deeply tied to the state of Texas, where Buckskin Joe Cullinan was one of the first to recognize and promote an enormous reserve of American oil.

Fires were a constant threat to workers dealing with a highly flammable substance such as petroleum. A single spark from two tools striking each other could ignite a blaze that might destroy millions of barrels of oil.

With no buyer on the scene, Andrew Mellon moved to protect his investment. He bought up more stock in Guffey Petroleum and began to squeeze Guffey out of power. Andrew arranged for William to serve as the company's executive vice president so that he could assume more control over the company's operations.

The Mellons proved adept at managing the company. Andrew persuaded Shell to tear up the oil contract that was driving Guffey Petroleum to ruin. William, meanwhile, cast about for new sources of oil to replace the dwindling wells in southeastern Texas. In 1905, when prospectors made a promising strike in what is now Oklahoma, the Mellons raced to take advantage of the opportunity.

From their experience with the Spindletop oil boom and bust, they knew well that a few days' delay could mean the difference between fortune and ruin. Standard Oil was already on the scene and working to construct a pipeline to transport the oil to its refineries. Not only did the Mellons buy up hundreds of oil leases around Tulsa, but they also worked frantically to beat Standard with their own 450-mile pipeline from Tulsa to the Gulf refinery in Port Arthur, Texas. With four separate crews working from both ends and from the middle, the Mellons won the race and got their oil on the market first. That brought their company to the top rank of the new oil companies rising out of the West to challenge Standard.

Standard's refusal to enter the Texas oil boom seriously injured its dominance of the industry. Its share of the refining market dropped from more than 90 percent in 1880 to 60 percent by 1911.

Determined to keep building the business, the Mellons got rid of Guffey altogether. In 1907, they reorganized Guffey Petroleum into the Gulf Oil Corporation, owned almost entirely by the Mellons.

Winners and Losers

The major players in the founding of Gulf Oil met various fates. Although the Mellons paid him one million dollars when they pushed him out of the company he had founded, Guffey remained bitter. He fought the Mellons in court for over 20 years for money he claimed they owed him. Although he won a judgment from the court, he eventually lost on an appeal. The long court fight drained him financially, and he was nearly penniless when he died. John Galey earned $375,000 for his part in the Spindletop oil discovery. He quickly spent it on further oil exploration, none of which was successful. Like his partner, Guffey, he died a poor man.

After the Mellons seized control of his oil company, James M. Guffey (1839-1930) complained bitterly, "I was throwed out."

Anthony Lucas fared better than his partners. He collected $400,000 from the Mellons, plus 1,000 shares of Gulf Oil stock. The money enabled him to move to Washington, D.C., where he established a career as a geologist and engineering consultant.

Pattillo Higgins, the man most responsible for the Spindletop strike, benefitted the least from it. Greed apparently got the best of Anthony Lucas, who ignored his agreement to give Higgins 10 percent of his earnings. Higgins had to sue Lucas to collect this relatively small amount. Higgins used his money to start his own small oil company in Texas but made only a modest profit before he sold out.

The Mellons, meanwhile, rode the rising fortunes of Gulf Oil to a position of incredible wealth. Yet, even though he had accumulated more money than all but a handful of people in the world, Andrew Mellon continued to scheme for more. After winning appointment as United States Secretary of the Treasury under Warren G. Harding in 1921, Mellon used his government position to rig tax and tariff laws to give rich Americans—like himself—special treatment. Texas congressman Wright Patman fumed, "Mr. Mellon has violated more laws, caused more human suffering and illegally acquired more property to satisfy his personal greed than any other person on earth." Mellon's conduct during his 12 years in the Treasury was so scandalous that Herbert Hoover had to reassign him to a position as an ambassador to save Mellon the embarrassment of being impeached by Congress.

A French newspaper once estimated that Andrew Mellon's riches could be made into 160,000 bricks of gold, enough to construct 52 solid-gold houses.

LEGACY

Of the hundreds of companies spawned by the great oil rush at Spindletop, most went bankrupt, and only a few lasted for more than a couple of years. Gulf Oil, however, emerged as a dominant player in the oil industry. Through a combination of keen foresight and aggressive exploration, it grew into an international giant. Gulf Oil set up the United States' first drive-in gasoline service station for automobiles in 1913. As consumer demand for the new method of transportation mushroomed, this corner stand became a prototype for what is now a familiar piece of American culture. Gulf also became a dominant player in oil exploration, not only in the oil fields of the Southwest, but also in international

Gulf Oil's first drive-in service station was built in Pittsburgh in 1913.

exploration. During the 1930s, the company took a leading role in the discovery and development of the world's largest pool of oil in Kuwait.

What was most remarkable about Gulf was that even at the height of its glory, it was a family operation. Since the breakup of Standard Oil, no single family, not even the Rockefellers, has completely controlled such a massive oil corporation as have the Mellons of Pittsburgh. By the late 1940s, Gulf Oil's stature in the business world moved William Mellon to remark, "The Gulf Corporation has grown so big I have lost track of it."

In the modern business world, however, even giants are engulfed by greater giants. In 1984, the former Standard Oil of California company, Chevron, bought Gulf for the incredible price of $13.2 billion. With that addition, the San Francisco-based Chevron has become the fourth largest oil company in the United States.

4

WILLIAM KNOX D'ARCY

BRITISH PETROLEUM: RICHES IN THE DESERT

Middle East. Oil. The words seem to go together like "white" goes with "snow." In fact, the sands of the Middle East conceal the largest supply of crude oil on the planet. Yet less than a century ago, there were no fabulously wealthy oil sheiks, no forests of oil derricks rising up over the desert, no fleets of oil tankers cruising the Persian Gulf. The region had not produced a single barrel of commercial oil, and few suspected that it ever would.

That all changed because of the daring and bulldog determination of British lawyer William Knox D'Arcy. D'Arcy had gambled once before on his luck at finding riches in the earth and had won big. At the turn of the twentieth century, he was ready to try again. Had he known the trials and despair this

William Knox D'Arcy (1849-1917) led the way in exploring and developing the world's largest source of crude oil.

roll of the dice would bring, however, he might never have taken on the challenge.

The Gold Mine

William Knox D'Arcy was born in Devonshire, England, in 1849, the only son of an Irish lawyer. After attending Westminster School, he emigrated with his family to the small town of Rockhampton in Queensland, Australia. There he studied law and eventually joined his father's legal firm. For a number of years, D'Arcy enjoyed his status as one of Rockhampton's most respected citizens and had plenty of time to devote to his hobby of raising and racing horses.

In 1882, three brothers appeared at his law office in need of help. They had just bought the old, crumbling Mount Morgan gold mine and decided they were in over their heads. Could D'Arcy find some investors to help them develop this mine? While working for the brothers, D'Arcy got gold fever himself. In 1886, he and some other investors took a chance and bought the Mount Morgan mine outright.

D'Arcy spent a good deal of money to get the old mine working. But the effort paid off even more handsomely than he had hoped. The mine was rich in gold, and it made D'Arcy a millionaire. He returned to England in the mid-1890s with his wife and five children, ready to live a life of leisure. He bought a house in London, a country mansion, a huge estate and game preserve, and a private box at the racetrack.

But after D'Arcy's astounding success with the gold mine, small bets at the track could hardly satisfy him. He cast about for some other risky adventure in the world of finance.

PERSIAN INTRIGUE

Meanwhile, Shah Muzaffar al-Din, the ruler of Persia (now Iran), had gotten himself deeply in debt with his lavish and irresponsible spending of his country's treasury. Looking for quick cash to solve the Shah's problem, Persian government official Antoine Kitabgi proposed selling oil rights to European investors.

Shah Muzaffar al-Din of Persia was eager to sell his country's oil rights because he needed "some ready money."

This offer took a lot of nerve in view of the fact that Persia had no proven oil reserves. Back in 1872, and again in 1889, Baron Julius de Reuter had paid for rights to drill for Persian oil. He based his hopes on the centuries-old existence of "fire temples"—continuously burning fires fueled by oil—in that region. He drilled three holes but never found oil.

Undaunted by this failure, Kitabgi produced a recent report from a French geologist who, after considerable research, had concluded that Persia could be sitting on a sea of oil. In 1900, Kitabgi approached an English diplomat to see if he was interested. Fearful that England's enemy, Russia, might be able to take over a bankrupt Persia, the diplomat was eager to help. Knowing D'Arcy's reputation as a gambler, he arranged for Kitabgi to meet the English investor in January 1901. Enticed by visions of the recent gusher at Spindletop, D'Arcy spent several months investigating the prospects of Persian oil. The reports were just intriguing enough to prompt D'Arcy into action.

Alfred Marriott

On April 16, 1901, D'Arcy's representative, Alfred Marriott, arrived in Persia to begin negotiating for the grant—known as a "concession"—that would give D'Arcy the right to drill for Persian oil. When the Russians heard that Persia was dealing with England, they tried to block the deal in a bidding war. Alerted by British ambassador Sir Arthur Hardinge that the Shah was desperate for ready money, Marriott finally clinched the deal by handing over 20,000 pounds in cash, along with 20,000 shares in the venture and 16 percent of the net profits of the oil development. For this price, D'Arcy won the oil rights to about three-fourths of the country—an area nearly twice the size of Texas—for a period of 60 years. The Shah signed the agreement on May 28, 1901.

> "The soil of Persia, whether it contains oil or not, has been strewn in late years with the wrecks of so many hopeful schemes of commercial and political regeneration that it would be rash to attempt to predict the future of this latest venture."
> —Sir Arthur Hardinge, the British ambassador to Persia, speaking of the D'Arcy concession

Hostile Country

D'Arcy had no clue what he was getting into. His success had come from winning a gamble with a gold mine. He had never organized a large expedition and knew nothing of engineering, oil, drilling, geography, or Persian culture. The area he chose for exploration, called Chiah Surkh, lay in northwest Persia, where the terrain was almost impassable. There were no roads, ports, or cities for hundreds of miles, and the population consisted of warring tribes that not only distrusted westerners but also refused to recognize the authority of the Shah or any agreement he might sign.

To direct the project, D'Arcy selected 50-year-old George Bernard Reynolds, a graduate of the Royal Indian Engineering College who had experience drilling for oil in Sumatra in the Far East. Reynolds had the energy and stubborn determination to keep going in the face of tremendous obstacles and the creativity to solve seemingly hopeless problems.

In September 1901, Reynolds started work with a crew made up primarily of Poles, Canadians, and local laborers. They had to transport each piece of drilling equipment more than 300 miles from the port of Basra on the Persian Gulf up the Tigris River to Baghdad, and then travel by mule and by foot across the plateau and over the rugged mountains to

George Bernard Reynolds (left), head of D'Arcy's drilling operations in Persia, takes a lunch break.

their primitive camp. They worked in suffocating heat (110 degrees Farenheit in the shade and 130 in the midday sun), battling outbreaks of smallpox. The water was barely drinkable, and their food supplies were plagued with locusts and plundered by tribesmen. Rain destroyed the mud-covered reed roofs under which they slept. When equipment broke down, as happened often in the heat and wind-blown grit, Reynolds could not afford to send for parts. Instead, he resorted to cobbling together parts from the equipment at hand. Occasionally, Reynolds had to pay bribes to hostile local authorities to keep them from raiding the camp.

In the face of all these obstacles, Reynolds kept his crew on task. At last, in November 1902, the crew began delving into the earth. The tremendous cost of transporting supplies meant that Reynolds only had funds to drill in one place at a time, and so the work went slowly. For month after discouraging month, they probed the rocky ground more than 2,000 feet without success.

D'Arcy's first well was drilled at Chiah Surkh, a remote plateau in the mountains of northwestern Persia.

On the Brink of Ruin

In October 1903, they finally tapped into a pocket of oil. Hopes soared, but quickly fell when the well produced only a trickle. By this time, the project had nearly ruined D'Arcy. He had originally estimated the cost of drilling two exploratory wells at about 10,000 pounds, but he had already spent 160,000 pounds with nothing solid to show for it. His bank account was overdrawn by almost 177,000 pounds, and his bankers had no intention of extending him any more credit.

Desperate, D'Arcy pleaded with the British government for a loan. When his request was refused, he had to borrow against some of his gold mine shares. This meant that if he could not repay the loan, his lavish living might come to an end.

In January 1904, D'Arcy received what he called "glorious news from Persia." Workers had struck a promising pool of oil. But again, the celebration was quickly cut short. The well ran dry within a couple of months without covering a fraction of the expedition's costs. D'Arcy had Reynolds pack up his 40 tons of drilling equipment and move to a different site in southwest Persia, but they had no better luck there. Even the dogged Reynolds almost had to concede it was time to give up and go home.

Luckily, a chain of events had already been set in motion that would prove to be the expedition's salvation. Fighting off depression, D'Arcy had taken a trip to a health spa in Bohemia in July 1903. There he ran into another depressed Englishman, Admiral John Fisher. Fisher believed that the British navy could increase its battleships' speed and efficiency by converting their engines to run on oil instead of coal. But in a recent test of oil versus coal, a faulty burner had caused the oil engine to blow out a huge cloud of black smoke, making Fisher look like a fool.

"If I wanted to be rich I would go in for oil! When a cargo steamer can save 78 percent in fuel and gain 30 percent in cargo space by the adoption of internal combustion propulsion and practically get rid of stokers and engineers—it is obvious what a prodigious change is at our doors with oil!"
—Admiral John Fisher to Winston Churchill, 1911

D'Arcy told Fisher what he was doing in Persia and confessed that he saw no choice but to abandon his quest for oil. Fisher, however, was unshaken in his belief that oil would revolutionize the navy of the future. He saw D'Arcy's Persian oil concession as a golden opportunity to secure a source of that oil for Great Britain.

Meanwhile, the British government had its own reasons to help D'Arcy keep his concession. A year earlier, he had met with the Rothschilds, a prestigious French banking family, to discuss selling the concession. This possibility alarmed the British. Even though they had little faith in the future of oil, they refused to surrender their foothold in Persia to France.

Fisher and others in the government pleaded with D'Arcy not to sell his concession. Instead, they put pressure on Burmah Oil, a large Scottish firm with oil interests in the Far East, to come to the aid of Great Britain by joining D'Arcy. Burmah had money to invest but was running low on oil sources. In May 1905, the company signed an agreement with D'Arcy, offering money and drilling expertise in exchange for a share of the oil.

The arrival of Burmah, however, only seemed to extend the agony. Reynolds, still in charge of the operations, drilled new test holes, but he was constantly harassed by brutal conditions and hostile local tribes. The British government sent a gunboat for protection, but the nearest river turned out to be too shallow for the boat to reach the drill site. When a widespread revolution in Persia further threatened the exploration, Great Britain finally sent a tiny force of 22 soldiers to guard Reynolds and his crew.

Time was running out on D'Arcy's gamble. By January 1908, he had sunk over 225,000 pounds into the venture. Burmah had lost 600,000 pounds and was losing patience. Reynolds decided to try one last spot, a place called Masjid-i-Suleiman, located in an area that Persians called the "Plain of Oil."

A Fortunate Postal Delay

The new site offered Reynolds a replay of all his previous drilling horrors plus a few more. His crew made a back-breaking effort to build a road through the desert only to see a river flooded by torrential rains wipe it out. They had to rebuild the road entirely. Reynolds was also forced to hire more guards because locals were throwing stones at the drillers.

For nearly five months, the drilling produced only the usual headaches. Burmah saw no point in continuing this fool's game. D'Arcy, still clinging to hope, stalled them for a few weeks. But in mid-May, Burmah's patience reached an end. The company

Building a road to the drilling site at Masjid-i-Suleiman

sent a cable to Reynolds telling him to "abandon operations, close down, and bring as much of the plant as is possible down to Mohammerah."

Reynolds figured that, with the wretched mail service, the letter confirming these instructions would not arrive for a couple of weeks. His crew worked day and night in a frantic effort to find oil before the letter arrived. At 4 A.M. on May 26, 1908, the crew heard a rumble from beneath the oil rig. Seconds later, oil blew 50 feet into the air. Oil was then struck in several surrounding wells, proving that Masjid-i-Suleiman harbored a reservoir at least 10 miles square. After more than seven years of exhausting and costly effort, Reynolds and D'Arcy had finally found what they were looking for.

THE GOVERNMENT MOVES IN

D'Arcy and Burmah formed the Anglo-Persian Oil Company to collect and refine the flood of Persian oil. While the company operated profitably for four years, the British government cast a longing eye toward its operations. Winston Churchill, who had recently become the First Lord of the Admiralty (the top non-military position in the Royal Navy), was committed to using oil to fuel the nation's battleships and saw oil as essential to the country's defense. He was reluctant to depend on private producers for a such an important product. "We must become the owners, or at any rate the controllers, at the source of at least a proportion of the supply of natural oil which we require," he declared in 1913.

The discovery well at Masjid-i-Suleiman. The first news of the Persian oil strike to reach England was sent by Lt. Arnold Wilson, one of the soldiers guarding Reynolds's crew. It was written in code: "See Psalm 104 verse 15 third sentence." This Bible passage reads, "that he may bring out of the earth oil to make a cheerful countenance."

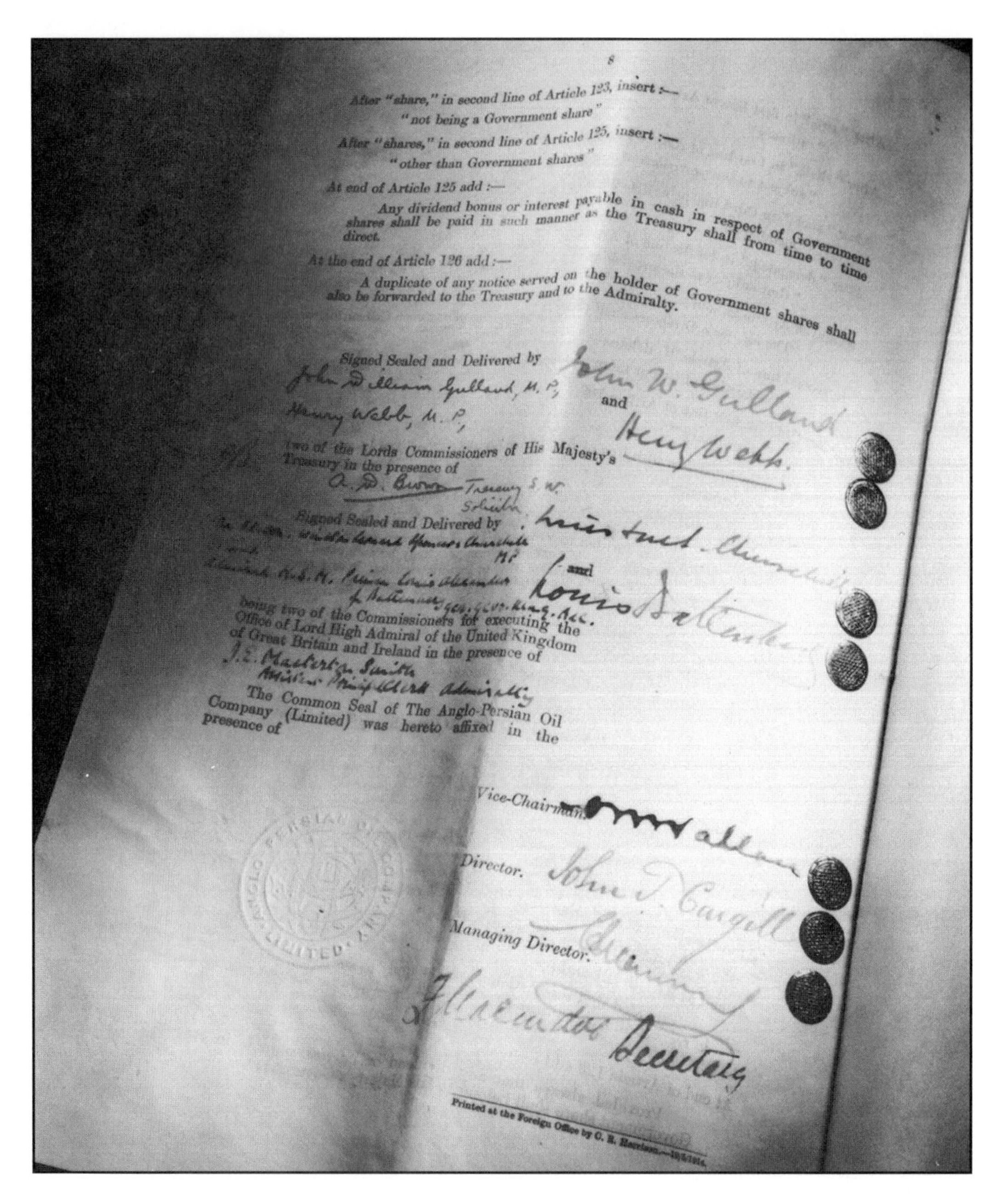

8

After "share," in second line of Article 123, insert :—

"not being a Government share"

After "shares," in second line of Article 125, insert :—

"other than Government shares"

At end of Article 125 add :—

Any dividend bonus or interest payable in cash in respect of Government shares shall be paid in such manner as the Treasury shall from time to time direct.

At the end of Article 126 add :—

A duplicate of any notice served on the holder of Government shares shall also be forwarded to the Treasury and to the Admiralty.

Signed Sealed and Delivered by John William Gulland, M.P., and Henry Webb, M.P., two of the Lords Commissioners of His Majesty's Treasury in the presence of

Signed Sealed and Delivered by and being two of the Commissioners for executing the Office of Lord High Admiral of the United Kingdom of Great Britain and Ireland in the presence of

The Common Seal of The Anglo-Persian Oil Company (Limited) was hereto affixed in the presence of

Vice-Chairman.

Director.

Managing Director.

Secretary

Printed at the Foreign Office by C. R. Harrison.—19/5/1914.

Signed by Winston Churchill and other officials, this 1914 agreement between the British government and the Anglo-Persian Oil Company greatly benefited both parties. At the time, Persian oil accounted for less than one percent of total world production. But after the outbreak of World War I, the demand for oil increased so dramatically that within two years Anglo-Persian was meeting one-fifth of the Royal Navy's oil needs.

Upon his urging, the British government paid two million pounds for a controlling share of the company, which was eventually renamed British Petroleum in 1954.

William Knox D'Arcy profited from his gamble, although not to the degree that he had hoped. He sold his concession to the Anglo-Persian Company in 1909, ending up with enough money to repay his 225,000-pound debts plus shares in the company worth almost a million pounds. He served on the Board of Directors at Anglo-Persian but, depressed at the way his great enterprise was slipping from his control, he eventually faded into obscurity until his death in 1917.

Like many of the men who performed the hardest work in the oil business, George Reynolds received the least profit. As thanks for his tremendous effort, company officials awarded him a total of 1,000 pounds, but only after they fired him for insubordination in 1910. Lieutenant Arnold Wilson, one of the soldiers sent to protect Reynolds's crew in Persia, offered this tribute: "The service rendered by G.B. Reynolds to the British empire and to British industry and to Persia was never recognized. The men whom he saved from the consequences of their own blindness became very rich, and were honored in their generation."

LEGACY

The Persian oil strike had an immediate impact on world affairs. As Churchill had predicted, oil became increasingly crucial to the military because of the

During World Wars I and II, strong diplomatic relations with the Middle East became important to many nations because of the enormous oil deposits there. In 1945, President Franklin D. Roosevelt (right) met with King Ibn Saud of Saudi Arabia (center) aboard the USS Quincy *in the Suez Canal to discuss issues such as the oil trade.*

emergence of gasoline and oil as the fuels of choice for all manner of transportation, including ships, tanks, airplanes, and trucks. Within a few years of the government's purchase of Anglo-Persian, Great Britain was embroiled in World War I, during which the dependable supply of oil played a key role in its victory. Twenty years later, Great Britain's control of a safe source of oil gave it a desperately needed advantage that helped it overcome the powerful armies of Nazi Germany.

British Petroleum profited greatly from its role in the war effort and continued to expand its operations. Not only did it retain its dominance in Persia (which was renamed Iran in 1935), but it helped to develop other oil-rich areas, such as Kuwait and Alaska. In 1998, British Petroleum joined forces with Amoco in the largest ever industrial merger. This move made BP Amoco the world's 19th largest company in 1999, with operations in 100 countries on six continents and revenues of over $68 billion.

Despite the success of the company William Knox D'Arcy helped build, in the long run his most important accomplishment was the opening of a previously untapped region to oil exploration. D'Arcy's success paved the way for western oil companies to descend upon the Middle East in search of oil. These companies discovered a supply so huge and valuable that the powerful nations of the world have gone to great lengths to secure access to it.

This scramble for oil in the Middle East has caused a build-up of powerful armies and intense political maneuvering that has inflamed the bitter, ancient hostilities that already existed in the region. Furthermore, the high-profile presence of European and American oil companies helped trigger a tide of resentment among local people against the foreign influences that they felt were exploiting their land and corrupting their society. Largely due to the influence of its oil, first unearthed by D'Arcy, the Middle East has reigned for several decades as the world's most dangerous political hotspot.

5

FRANK PHILLIPS

PHILLIPS PETROLEUM'S WILDCAT KING

While working as a barber in a small Iowa town, Frank Phillips sympathized with men who complained of hair loss. Although he was only in his mid-20s, Phillips himself was already thinning out on top. Considering the problem, he noted that the pigs on his father's farm never went bald. How were they able to keep their hair?

Phillips observed that pigs stood out in the rain more often than people did. Rainwater, he decided, was the key to healthy, growing hair. Phillips bottled and sold scented rainwater with the claim that it could cure baldness. At first, his product sold briskly. But before long, customers found that it simply did not work. With his own shiny, bald dome growing ever larger, Phillips could hardly claim otherwise.

Frank Phillips (1873-1950), founder of the company that developed the famous Phillips 66 gasoline, had this portrait painted on his 66th birthday.

Frank Phillips was not a man who relied on exhaustive study and research in his business dealings. Instead, he was a gambler who played his hunches. Some of them, like his rainwater cure, never paid off. But when it came to the oil industry, Phillips's instincts were astounding. He built his business on wildcatting—drilling wells where no oil had been proven to exist. Phillips was so good at guessing where to find oil that he became known as the "King of the Wildcatters," and he played his hunches into one of the most profitable oil companies in the world.

The Nebraska Disaster

Frank Phillips was born on November 28, 1873, in Scotia, Nebraska, the third child of Lewis and Lucinda Phillips. His parents were farmers who had moved from Iowa to central Nebraska a year earlier to take advantage of cheap land. But after a plague of locusts chewed their entire crop down to the roots, the Phillips family abandoned Nebraska. In 1874 they moved to Taylor County in southwest Iowa, where Lewis supplemented his farming income by working as a carpenter.

Frank and the other eight surviving Phillips children worked hard on the farm for most of the year. During the autumn and winter, they took time off to attend a one-room school. Although Frank was not very interested in school, he became enthralled by a series of popular adventure stories by Horatio Alger. The heroes of these books were boys who used their wits and determination to rise from poor origins to fame and fortune. As he read, Frank dreamed of following in their footsteps.

Mr. Fancy Pants

Finances were always tight for the Phillips family, and at the age of 12, Frank had to quit school to work as a hired hand on neighboring farms. Two years later, on a visit to the nearby town of Creston, he saw a barber wearing a pair of fancy striped pants. At that moment, Frank later recalled, "I made up my mind that I wanted to earn enough money so that I could afford to wear striped pants even on weekdays." He confided his ambition to a family friend who was a barber. Knowing that Frank was a hard worker, the friend hired him as an errand boy and allowed him to gradually learn the trade.

Phillips (far right) began his career as a barber in Creston, Iowa. This barbershop, advertised as "The Oldest Continuous Barbershop in Iowa," is still in business today.

When he turned 17, Phillips decided he knew enough about barbering to strike off on his own. Seeking adventure and fortune, he moved out to Colorado and settled in the thriving silver mining center of Aspen, where he worked at the Silver Dollar Barber Shop. Good times in Aspen screeched to a halt in 1893, however, when the U.S. government switched from silver to gold to back its currency. The silver market plummeted, and Aspen's economy collapsed. Customers moved away from Aspen, so Phillips headed west to Ogden in the Utah Territory.

Before long, Phillips tired of eking out a living as a part-time barber and railroad brakeman. In 1895, he wired his family for money to return to Creston. Apparently embarrassed by his lack of success, he never spoke about his years in Aspen or Ogden.

Phillips had better luck as a barber in Creston. He worked at and eventually purchased the Climax Barber Shop, turning it into the fanciest barbershop in town. An enterprising businessman, he paid his barbers per haircut so they would have an incentive to go out and recruit customers. Soon, he was able to open a second shop. Frank Phillips was finally his own boss, and every day of the week he wore the striped pants he had so long admired.

John Gibson

An Offer He Couldn't Refuse

One of the regular customers at the Climax Barber Shop was John Gibson, president of the Iowa State Savings Bank. Phillips started courting Gibson's daughter Jane, which worried the banker. He liked Phillips, but thought his daughter could do better than a barber for a husband. When the couple wanted to get married, Gibson approved on one condition: Frank would have to give up barbering and enter the banking business. Phillips agreed, and the marriage took place on February 18, 1897, with the couple receiving a wedding gift of $20,000 from Gibson in the bargain.

For a short time, Phillips continued to run his barbershop while learning banking on the side. On his first major assignment, he showed a flair for sales. Gibson assigned him the job of selling bonds to finance a new sports coliseum in Chicago. Phillips sold most of the bonds in less than six months, earning $75,000 in commissions. When the coliseum opened in 1900, Frank took charge of its ticket sales.

Frank Phillips and Jane Gibson on their wedding day in 1897

The Oil Gamble

Phillips might have continued as a salesman had he not visited St. Louis in 1903 to check out preparations for the World's Fair to be held there the next year. There he ran into an old friend, the Reverend C. B. Larrabee, a Methodist minister who had served for a time in Creston. Larrabee described a scene he had just left in the town of Bartlesville, in what is now Oklahoma. Oil had just been discovered in the region and, according to Larrabee, it was "flowing out of the ground like water."

Phillips checked out Bartlesville and liked what he saw. He knew nothing about the oil business, but he had a hunch that oil would play a big role in society. When he returned to Iowa, he called on his younger brother, Lee Eldas (nicknamed L.E.). Frank pointed to an automobile chugging down the street. "You see that contraption down there? I think people are going to buy quite a few of these new buggies and they need gasoline to make 'em go."

L.E. was skeptical. After all, hardly anyone he knew owned an automobile. But Frank finally talked him into joining his new oil adventure. He also convinced John Gibson and a number of others to help him organize the Anchor Oil and Gas Company. Phillips made two more visits to the Bartlesville area in 1904 to buy oil leases. Knowing nothing of geology, he simply chose land that was as close as he could get to a producing well.

In 1905, Phillips moved his family to Bartlesville and hired a drilling crew. Phillips struck oil on his first try, but the well quickly dried up. The second and third wells he drilled were dry holes, which drillers called "dusters." L.E., who was in serious financial trouble, wanted to quit before the effort ruined him completely. But Frank decided to risk all his remaining money on a fourth well.

Late that summer, his crew set to work in the tall grass about seven miles north of Bartlesville. On September 6, Phillips struck a 250-barrel-a-day gusher. Encouraged, Phillips kept drilling on his leases as fast as he could. Although every one was a wildcat, Phillips drilled 81 wells without hitting a single duster.

The Outlaw Bank

Even with this success, Phillips realized that his luck could go sour at any time. He saw an opportunity to

Phillips's first gusher, the Anna Anderson #1, was named after the 8-year-old Delaware Indian girl who owned the land on which the well was drilled. In Oklahoma Territory, tribes could provide a parcel of land to any individual member who wanted to lease it for mining or drilling, and Anna's grandfather had obtained land in her name as well as his own. When Phillips struck oil, she became the richest Native American girl in Oklahoma.

reduce his dependence on oil by returning to his father-in-law's profession—banking. Since local bankers were exceptionally cautious about lending to oil drillers, the field was wide open to someone willing to take risks. In July 1905, he and L.E. organized the Citizens Bank and Trust Company.

Phillips wisely gained the trust of townsfolk by using local workers and materials to build his bank. He also attracted oil men, who were frustrated by other bankers' unwillingness to lend. When Citizens Bank opened on December 5, it brought in $80,000 in deposits. Even more impressive was that in a region where bank robberies were a daily occurrence, Phillips's bank escaped unscathed. Stories circulated that outlaws stole from other banks and deposited their ill-gotten gains with Phillips.

Phillips's fair treatment of all his bank customers (regardless of their criminal records) impressed many Oklahoma outlaws—including bank and train robber Henry Wells, who soon became a good friend. In 1933, the notorious gangster Charles "Pretty Boy" Floyd confided to Wells that he planned to kidnap Phillips or a member of his family. Wells warned Phillips immediately, preventing the attempt. "It was the only time that I ever ratted on any of the boys," Wells recalled. "But I told Pretty Boy that I'd rather somebody would have kidnapped my daddy than Frank Phillips."

Whether this was a legend or the truth, Phillips's clients did include a number of notorious criminals.

The oil near Bartlesville was on Indian land, which was governed by a complex set of federal laws. One of these specified that no company could own more than 4,800 acres of land. Phillips got around this regulation by creating 15 different corporations to buy oil leases. He named these companies after family, ancestors, friends, and towns in which he lived. For example, the Lewcinda Oil and Gas Company was a combination of his parents' names. With the help of his brothers, L.E. and Waite, Frank ran all his oil operations from a room in the rear of his bank.

For 10 years, Frank Phillips kept up a whirlwind of activity as he sought to increase his business holdings. He rarely took vacations or spent time with his wife and son. When he had a spare moment, he enjoyed playing poker with other high stakes oil gamblers such as Bill Skelly and Harry Sinclair. He once cut cards with a friend for $10,000 and won. The gambling addiction passed on to his son, John, who once bet his house in a card game and lost.

As an employer, Phillips was fair and compassionate but firm. He paid employees who missed work because of family illnesses. On the other hand, when he found his younger brother, Fred, squirrel hunting when he should have been working, he fired Fred on the spot. After years of John's alcoholism and poor performance, Phillips even fired his own son. On one especially memorable occasion, John, assigned to escort a group of company directors, had gotten drunk and left them stranded in Kansas City.

Oil Men or Bankers?

By the end of 1914, the Phillips brothers' oil and banking businesses had grown so large that they did not have time to run them both. Although the oil leases were making money, Frank kept worrying that his luck would run out. He knew the odds were against his continued success. The boom and bust cycle of the oil industry had ruined far more people than it had helped. Furthermore, his brother Waite had just formed his own oil company, and Frank did not relish the family competition. In 1915, he decided to get out of the oil business while he was ahead. "We're not oil men, we're bankers," he admitted to L.E. The two made plans to sell all their oil interests and buy a bank in Kansas City.

But the federal regulations governing the use of Indian land put a snag in the plan. Phillips spent months trying to put together the paperwork necessary to sell his interests. In the meantime, the war in Europe was threatening to draw in the United States. Phillips saw that modern warfare required millions of gallons of fuel to transport troops and equipment.

"We're not bankers, we're oil men," he suddenly declared. L.E. was always far more comfortable in banking than in oil, but he followed his brother's wishes, as usual.

L.E. (left) and Frank Phillips at Frank's ranch, Woolaroc

Close Call

By this time, however, the most promising oil land in Oklahoma had already been grabbed by prospectors. In 1916, Phillips was running out of leases. If he expected to keep building his oil business, he would have to strike a major oil source and do so quickly. This time, Phillips's luck appeared to have run out. He drilled four holes on his last major lease, each as dry as the last. Not wanting to give up, he drilled another one. It, too, was a duster. He tried one more. Nothing.

When Frank Phillips had a hunch, though, he stayed with it. He kept drilling and on February 13, 1917, he hit a small oil pocket. Encouraged, Phillips tried another well close by. On March 22, he stood on the derrick platform, watching as the drill sank down to 1,800 feet. Suddenly, he heard a rumbling from below. An alert driller grabbed Phillips and dove off the platform just as it was shattered by an explosion of oil that sent debris flying everywhere.

This turned out to be the largest oil well yet found in the area. The gusher came in at the right time, too—just as the United States entered World War I. With the new demand for oil, the price per barrel soared from 40 cents to more than $1. Frank Phillips was now incredibly rich.

Much of the land that Phillips drilled on so successfully belonged to the Osage Indian tribe, and his wells not only built his own fortune but made the tribe the richest in the world as well. The revenues from the tribe's mineral rights were divided equally among its members, and by 1923 each Osage man, woman and child was earning $12,400 a year in oil money. As recognition for Phillips's friendship, the Osage adopted him into their tribe in 1930—the first time a white man had been so honored.

Phillips Petroleum

Seeing a golden chance to boost his company into an industry giant, Phillips incorporated all his businesses into the Phillips Petroleum Company, which was opened to public investors late in 1917. Frank became the company's president, and L.E. served as vice president.

The oil and the money kept rolling in, and Phillips kept putting much of his earnings into new leases and exploration. In 1919, Phillips Petroleum had 450 oil wells and produced less than 6,000 barrels a day. Seven years later, the company owned 2,300 wells, producing 55,000 barrels of oil per day.

Meanwhile, Phillips's prediction about the future growth of the automotive industry had come true. Back in 1910, there had been fewer than half a million automobiles in the United States. Ten years later, more than nine million cars were on the roads. Seeing the increased demand for gasoline as a golden opportunity, Phillips decided to branch out to refining and selling finished oil products.

In 1925, however, Union Carbide and Carbon Company sued Phillips Petroleum for violating one of their refining patents. Furious, Frank Phillips pumped more money than any of his competitors into research and development. He won the lawsuit when his scientists proved the Union Carbide patent was invalid because that specific refining process had existed since ancient Egypt. The scientists then went on to improve Phillips's products, and the company became a leader in the development of aviation fuel.

Built in 1925 to house the Phillips Petroleum Company offices, the Frank Phillips Building dominated the skyline of downtown Bartlesville.

PHILLIPS 66

In the mid-1920s, Phillips was selling its products to other refiners, who sold them under their own names to consumers. Frank grew irritated that his company was making what he considered the finest gasoline in the world and yet most consumers who used it had never heard of Phillips Petroleum. Phillips decided he had to get into the retail business himself.

Flush with success from another spectacular oil find, this time in the Texas Panhandle, Phillips set his researchers to work developing a new brand of gasoline that would out-perform the competitors. The scientists analyzed all the gasolines on the market and tried different blends of gasolines in every existing brand of car. Finally, they came up with a unique blend of Phillips gasoline that made cars run more smoothly and start easily in cold weather.

False stories about the origin of the name "Phillips 66" circulated for years. People said that the "66" stood for Phillips's age when he founded the company (he was actually 43), the year of his parents' marriage, the size of his ranch, or even the number of books in the Bible. Others thought that Frank and L.E. had only $66 left when they first struck oil or that an executive had won the company's first refinery by rolling double sixes in a dice game. The theories became so outlandish that Phillips Petroleum finally published a booklet to set the record straight.

Phillips wanted a flashy name for the product to catch consumers' eyes. One day, a Phillips test driver and a company executive were testing the gasoline formula on a new stretch of the country's first interstate highway. The new fuel improved the car's acceleration so much that they were soon traveling at almost 66 miles per hour. Suddenly, they had a flash of insight. Why not name the gasoline after this famous Chicago-to-Los Angeles highway, Route 66?

On November 19, 1927, the first service station selling Phillips 66 opened. Phillips had chosen to build it in Wichita, Kansas, the toughest, most competitive gasoline market in the Midwest. He believed that if he could succeed there, he could succeed anywhere. The building resembled a cottage to blend in with the residential neighborhood in which it was located. Phillips gave out coupons for five free gallons of gasoline. By the next day, the station had not only pumped over 24,000 gallons of gasoline but won hundreds of loyal customers.

Phillips moved quickly to establish the company in other markets. By the end of 1928, Phillips 66 had 1,800 retail outlets. Within three years of opening

This Phillips 66 station, the first ever built, had so many customers that by the afternoon of its second day in operation it had to close to await a new shipment of gasoline.

its first service station, Phillips built or purchased 6,750 service stations in 12 states. Its new logo, a 66 on a shield that resembled national highway signs, became a familiar sight throughout the Southwest.

In 1930, Frank took another daring gamble by constructing a 680-mile gasoline pipeline—the world's longest at that time—to distribute his product more efficiently. The project's enormous cost, combined with the economic crunch of the Great Depression, placed a staggering load of debt on the company. Many of its directors and investors demanded that the project stop. Frank promptly wrote a letter of resignation. Unwilling to lose the man who had led the company since its beginning, the protesters backed off. The pipeline, completed in 1931, helped Phillips Petroleum expand into 10 more states. By 1940, the business was worth $226 million and employed over 30,000 people.

Frank Phillips remained active in the company he founded until 1949. One of the great ironies of the oil industry is that Phillips, who profited so much from the production of oil and the sale of gasoline, died on August 23, 1950, at the age of 76 without ever learning to drive an automobile.

Legacy

Phillips Petroleum has had an astonishing influence on both the oil industry and on everyday life in the United States. Its explosive entry into the retail gasoline market was so successful that it forced other major oil companies to follow its lead. Prior to 1927, most gasoline stations were independently run, like the old family-owned grocery stores. Once Phillips made its move, Texaco, Royal Dutch/Shell, and other major producers planted service stations on street corners throughout the nation. Phillips continued to expand its retail operations, and by 1967 it had a total of 23,400 stations throughout all 50 states.

Phillips, driven partially by outrage over Union Carbine's patent lawsuit, developed a sophisticated research and development department. This effort made the company a leader in the creation of new products such as aviation fuels and liquid petroleum (LP) gas, used for cooking and heating. Phillips's scientists in the 1950s had great success in developing new uses for plastic products made from oil.

offshore drilling: drilling for oil that lies under a body of water. Most offshore drilling is done in seas or oceans, but oil deposits have also been found in large lakes.

In the tradition of its founder, Phillips Petroleum took a huge gamble on an offshore drilling venture in the Norwegian North Sea during the late 1960s. The company spent millions of dollars drilling for oil

This drilling rig, built to exploit Phillips Petroleum's discovery of the Ekofisk oil field in the Norwegian North Sea, is the largest offshore complex in the world.

as far as four miles below the ocean floor. Battling high winds and 50-foot waves that nearly capsized the drilling rig, Phillips crews drilled one dry hole after another. With nothing to show for its enormous investment, the company finally called off the search. Phillips made one last try, however, only because it had already paid for the equipment and could not get a refund. This probe struck oil in 1969, leading to a $6.7 billion project to exploit what became a major oil field.

Phillips Petroleum topped $1 billion in sales back in 1956 and has shown steady growth ever since. The company built on the hunches of a small town barber has grown into the eighth largest oil company in the United States.

6

J. Paul Getty

GETTY OIL: ASHES TO ASHES

When Jean Paul Getty's father died, he left the bulk of his wealth to his wife and gave half a million dollars to his only child. Instead of being grateful for the gift, Jean Paul felt cheated. Scheming for a way to get his hands on more of the money, he got his mother to agree to a complicated reorganization of his father's business interests. When the deal was complete, he privately gloated, "I just fleeced my mother."

J. Paul Getty knew how to make money. He single-handedly wheeled and dealed his way to the greatest private fortune in the world. But his behavior might have shocked even the most reviled oil tycoons that preceded him.

Unlike Rockefeller, Getty rarely gave a penny to charity. He was so tight with his money that he

Jean Paul Getty (1892-1976) established an international reputation as an eccentric billionaire and cutthroat businessman.

> "Getty's character was a mass of contradictions and inconsistencies. He could throw a $28,000 party, and still put a pay telephone in his house. . . . He would insist on not breaking the law in business, but seemed to be totally unaware of the codes that rule our personal lives. He could be more daring than most other entrepreneurs, but was obsessed by fears."
> —Robert Lenzner, *The Great Getty*

installed a pay phone for guests to use at his house. He married and divorced five times, not always waiting for divorce before remarrying. He left his wives to fend for themselves; in one case, he didn't see his wife for years following their wedding day. When his 12-year-old son lay dying with a brain tumor, Getty refused to visit him. When another son wrote him a letter, Getty marked the grammatical and spelling errors and sent the letter back without comment. The son never wrote to him again. Later in life, Getty changed his will 21 times in 18 years to keep his family and friends guessing about who would inherit his money.

Honest George and the Stifled Son

Getty's parents certainly did not raise him to be that way. His father, George, was a lawyer who carried a lifelong reputation for fairness and honesty. George Getty's word was better than a contract.

But while Jean Paul came from a respectable, decent family, he was raised in a stifling atmosphere. His mother, Sarah, was 40 years old when he was born on December 15, 1892, in Minneapolis, Minnesota. She was frail from a near-fatal bout with typhoid fever and from grief over the death of her daughter two years earlier. Jean Paul never had a birthday party or a Christmas tree. His parents refused to let him play outdoor games with other boys for fear he might get hurt. They moved six times in the Minneapolis area during Jean Paul's first 13 years, eliminating any chance he might have had to develop friendships. He grew up a loner, with his dog, Jip, his only companion.

Oil Fever

George Getty worked his way up to chief lawyer and director of the Northwestern National Life Insurance Company. In 1903, he traveled to Bartlesville, Oklahoma, to collect a $2,500 debt that some oil drillers owed his company. While in Bartlesville, he got so caught up in the frenzy of oil exploration that he bought some oil leases. He formed the Minnehoma Oil Company, which struck oil on his leased land in January 1904.

Oil wells in Bartlesville. By 1920, Oklahoma was producing over 100 million barrels of oil per year, more than Texas or California.

While living in Bartlesville, J. Paul attended the brand-new Garfield School, shown here shortly after its construction in 1904.

Sarah and Jean Paul joined him in Bartlesville. Still friendless, Jean Paul spent much of his time reading. He earned money by selling the *Saturday Evening Post* magazine and invested his earnings in the family company. Visiting his father's wells often, he learned about the business and became fascinated by the excitement of drilling for oil.

For a time, George Getty was Oklahoma's largest independent oilman. When Jean Paul was 14, however, the family moved to Los Angeles, hoping the California climate would be better for Sarah's health. In high school, Jean Paul was a poor student and a discipline problem. His parents sent him to a military school to shape him up, but he did not improve. In his spring semester at the University of California-Berkeley in 1912, he did not complete a single course.

Despite Jean Paul's poor record, his father's money continued to open doors for him, and he entered England's Oxford University later that year. But when he wasted the time wandering in Europe instead, his frustrated father took control of his car and oil stock. Jean Paul never forgave his father for that, although George continued to give his son generous financial support. Somehow Jean Paul received a diploma from Oxford, and he returned to the United States in 1914.

Lucky Strike

Determined to make money on his own, J. Paul, as he became known, drove to the Oklahoma countryside in a beat-up Model T. In the past, he had spent summers working on his father's drilling crews, but now he began scouting for promising oil leases. He spent over a year looking for a lease he could afford and finally decided the best deal was a plot about 30 miles south of Tulsa. Getty then secretly convinced the vice president of the town bank to bid on the land for him. Although the other oil companies could easily out-bid Getty, they might assume that the prominent banker was acting on behalf of a major corporation with unlimited funds. Sure enough, the other companies were intimidated and dropped their bids. Getty acquired the lease for only $500.

Luck stayed with him, and the very first hole he drilled on the land struck a 700-barrel-per-day winner. Three days after the discovery, he sold his share of the lease for a profit of almost $12,000. George

Getty was so proud of his son that he made him a director of Minnehoma Oil.

J. Paul stayed in Oklahoma, living in a hotel while buying and selling oil leases. He discovered that his knowledge of geology, gained from books, gave him a great advantage over his competitors in finding oil. Because he knew that oil often lay in certain rock structures, he studied the rocks before making decisions on leases.

Between February and June 1916, Getty wildcatted and traded furiously. Taking skillful advantage of the rising price of oil due to World War I, he earned a fortune for himself and for his father's company. He decided to enjoy life as a millionaire, and retired to California at the age of 23.

The Best Room in the Best Hotel

Living a life without responsibilities, Getty quickly got himself into trouble. A woman sued him, claiming he was the father of her child. This very public scandal, which he settled out of court, established a pattern of behavior that he would follow all his life.

In 1923, Getty married for the first time. The marriage lasted two years. In 1926, while supervising oil exploration in Mexico, he met and married a 17-year-old. Within two years, Getty divorced again, traveled to Europe, and married an Austrian teenager. That marriage would last another three years before Getty married again, this time before his divorce from his third wife had been made final.

Meanwhile, George Getty was learning that J. Paul was not a safe partner, even for a father. "There's always the best hotel in town and the best room in the best hotel in town, and there's always somebody in it," J. Paul said. His goal was to be that person, and he could not stand to lose. If his family stood in the way of his goal, J. Paul would run over the family. In March 1923, shortly after Minnehoma made a huge profit on its oil discoveries on the southern California coast, George suffered a stroke. J. Paul saw this as an opportunity to take over the company. He was well on his way to doing so when George recovered and stopped him.

Making Money in the Depression

When his father died in 1930, J. Paul was enraged to discover that although he had inherited half a million dollars, he had not gotten what he really wanted:

Oil wells on the California coast. In the 1890s, major oil fields were discovered in Los Angeles and the San Joaquin valley. Then, in 1910, Charlie "Dry Hole" Woods struck a well that, at its peak, produced 80,000 barrels daily. This discovery, the largest well in the United States at the time, launched the booming California oil industry. That year, California made up a larger percentage of total world oil production than any foreign nation.

control of the family business. Four years earlier, George had secretly altered his will to punish J. Paul for his failed marriages, leaving the bulk of George F. Getty, Inc., to Sarah. Although J. Paul became president, treasurer, and general manager of the company (which included Minnehoma), his mother still held ultimate authority as the majority stockholder. Determined to challenge his father's decision, J. Paul set out to prove his worth by making more money.

Although Getty became famous as a businessman, he was first and foremost a wildcatter who did much of his best work in the field. Once, when his crew's drill bit broke off and lodged 4,000 feet underground, Getty had to come up with a way to clear the bit from the well so drilling could continue. Desperate, he drove to a Los Angeles cemetery and bought a new gravestone from the stonemason who worked there. The crew was bewildered when Getty returned to the drill site with a six-foot marble pillar in the backseat of his car and told them to throw it down the hole, but they obeyed. The weight of the pillar smashed the drill bit free, and soon oilmen all over California were using stones they called "Paul Getty Specials" to deal with lost drill bits.

In 1929, a stock market crash had triggered the Great Depression, and businesses and banks were failing at an alarming rate. Like Rockefeller, Getty saw bad times as an excellent opportunity to acquire wealth. Betting that the economy would eventually bounce back, Getty wanted to buy thousands of shares of oil stock that investors were eagerly selling at low prices. The Getty board of directors, however, led by J. Paul's conservative mother, saw no point in buying stock in other companies, especially when the economy was a shambles. In 1931, after a series of heated arguments and an investment of his own money, Getty finally got control of Pacific Western, one of California's largest oil companies. Then he bought a considerable number of shares of Tidewater Oil, the ninth largest oil company in the United States. Getty even snapped up other businesses at fire-sale prices, including a hotel and an aircraft manufacturer.

As the nation's economy continued to collapse, the Getty company lost millions of dollars on these purchases within a few months. But J. Paul cut costs by firing all his employees and hiring them back at lower wages. Meanwhile, he patiently waited for the economy to rebound.

Always on the lookout for a good oil bargain, Getty sent a representative to Iraq in 1932 to make an offer to buy rights to Iraqi oil. The price turned out to be reasonable. But with oil prices down, his cash running low, and an unstable political situation in Iraq, Getty was unwilling or unable to pay as much as the Iraqi government wanted. As the value of the Iraqi oil concession skyrocketed over the next few years, Getty kicked himself for what he considered the biggest mistake of his business career. He would learn from that mistake.

Getty at his desk, signing a document. In the 1930s, Getty pioneered his technique of purchasing stock in companies larger than his own, then eventually gaining control of them.

World War II

In 1934, Getty gained full leadership of the family company and, as his wealth and influence grew, he began to set his sights on a prestigious government post. He tried to use large political contributions as leverage to persuade President Franklin D. Roosevelt to appoint him as an ambassador. Coming on the heels of bad publicity over Getty's fourth divorce, however, the campaign never had a chance.

Getty also tried to get himself commissioned as a naval officer. Noting, however, that Getty did business with a number of Germans and had expressed admiration for Hitler's government, the FBI suspected him of being a Nazi sympathizer and even a spy. Although a three-year investigation failed to find evidence to support these suspicions, Getty did not get his commission.

Given his paranoia, military command could have been disastrous for Getty. He kept a yacht off the California coast so that he could escape in case Communists took over the country, and he lived in a concrete bunker during World War II for fear of German bombers. But he worked hard for the war effort, concentrating his efforts on cranking out military aircraft as fast as his Tulsa-based Spartan Aircraft Company could make them.

The Neutral Zone

At the end of the war, Getty saved thousands of jobs by converting the Spartan Aircraft factory to manufacture mobile homes instead of closing it when the demand for planes dropped. He also renewed his long quest to get complete control over Tidewater Oil. When that failed, Getty arranged to sell all his oil holdings. But he arranged the terms of the deal to favor himself at the expense of his family and partners. Irate, they blocked the sale.

> "I believe that the able industrial leader who creates wealth and employment is as worthy of historical notice as the politician or the soldier who spends an evergrowing share of the wealth created by individual initiative and courage."
> —J. Paul Getty

Frustrated, Getty put all his energy into creating a worldwide oil empire. In order to do that, he had to break into the Middle East. Near the end of the 1940s, a new concession went up for bid: 2,200 square miles of desert known as the Neutral Zone, which lay between Saudi Arabia and Kuwait. Since Kuwait was the site of the largest oil field in the world, the area held great promise as an oil source.

The two countries had fought over this land for centuries before finally agreeing to compromise by establishing joint rule of the area. Sheik Ahmad, ruler of Kuwait, sold his country's share of Neutral

Zone rights to Aminoil, a joint venture of independent oil companies that included Phillips and Ashland. But no one could drill on the land until King Ibn Saud of Saudi Arabia sold his share of the rights.

Getty sent geologist Paul Walton to survey the Neutral Zone from an airplane. Walton saw a small mound rising out of a flat desert plain, almost exactly like the terrain in which Kuwait's oil field lay. Based on that, Getty entered into secret negotiations for the oil rights. Still smarting over his failure to win the Iraqi concession, he was determined to grab the Neutral Zone rights at all costs.

On December 31, 1948, his Pacific Western Oil Company signed an agreement that shocked the oil industry. In exchange for a 60-year claim to Saudi Arabia's share of the Neutral Zone oil, Getty agreed to pay $9.5 million up front, plus $1 million per year

Having prepared for the meeting by listening to "Teach Yourself Arabic" records, Getty discusses the Neutral Zone oil concession with King Ibn Saud of Saudi Arabia.

Other oil companies were furious with Getty for driving up the price of oil concessions. A few years earlier, the oil rights for all of Kuwait sold for only $170,000, and Standard Oil had paid just $200,000 for 440,000 square miles of Saudi Arabia. Getty, however, believed the best way to do business was to give nations a fair share of the profits from their land's resources. As he bluntly put it, "When I go to the Middle East I don't want to feel I have to run down an alley every time I see an Arab approaching."

whether he found oil or not. In addition, Getty would pay a royalty of 55 cents per barrel produced, a much higher percentage than anyone else in the Middle East was paying, plus 100,000 gallons of gasoline and 50,000 gallons of kerosene for the Saudi government. Furthermore, Getty would build a refinery in Saudi Arabia and provide housing, schools, training programs, a water supply, and other community improvements for employees. Many oil industry experts thought Getty was crazy. Getty admitted that he might have paid more than the oil was worth now. But he reckoned that oil would increase in value. Before long, his deal would look like a bargain—if there was any oil on the land.

Initially, Getty agreed to let Aminoil direct the exploration. But as the drilling progressed slowly with no success from 1950 to 1952, Getty grew impatient. By 1953, both Aminoil and Getty had spent $30 million with nothing to show for it. Aminoil's leadership was quarreling heatedly with Getty and ready to quit. But Getty believed that Aminoil had drilled too low on that mound where he had expected to find oil. He insisted they try drilling another well near the top of the dome.

Drilling began on February 10, 1953, and continued more than half a mile underground without success. But in March, at a depth of 3,482 feet, drillers struck an enormous oil field. Getty used the flood of oil from that field—estimated at over 13 billion barrels—to establish himself as a giant in the industry. In a single month, the price of Pacific Western's stock doubled. The company, which Getty renamed Getty Oil in 1956, became the nation's seventh largest gas marketer by the end of the decade.

Meanwhile, Getty racked up an immense personal fortune. The Neutral Zone oil catapulted Getty Oil from a net worth estimated at $80 million in 1948 to well into the hundreds of millions. In 1957, a survey by *Fortune* magazine reported that J. Paul Getty was the richest man in the United States and a billionaire twice over.

The kidnapping of 17-year-old Jean Paul Getty III (right, with his fiancee) became a very public example of Getty's stinginess and his poor relationship with his family. Even after his grandson had been missing for three months, Getty refused to pay a ransom. Believing the kidnapping might be a fraud, he said, "I have 14 other grandchildren and if I pay one penny now, then I will have 14 kidnapped grandchildren." It was not until the kidnappers cut off Paul III's right ear and mailed it to an Italian newspaper that Getty decided the threat was real. He agreed to pay a million-dollar ransom, and Paul III was released.

Although he was proud of his wealth, the sudden publicity upset Getty. Before the article appeared he had already been paranoid about the prospect of death. After 1951, he never again visited his native United States, primarily for fear of crossing the Atlantic Ocean. Rich as he was, he hand-washed his underwear every night because he didn't trust commercial detergents. Now he saw himself a target of kidnappers, a fear reinforced when his grandson J. Paul III was abducted and held for ransom in 1973. For the last two decades of his life Getty holed up in a 72-room mansion in England, its fenced grounds patrolled by dozens of vicious attack dogs.

From his fortress, Getty continued to build his oil empire. He invested $600 million in refineries and transportation facilities, including more than $200 million on the world's largest supertankers. He moved his oil exploration teams into New Zealand. During the 1950s, his 15 wells there pumped out more oil than his 556 wells in the United States. By 1965, Getty Oil, of which J. Paul owned 80 percent, was worth more than $1 billion. He made even more money in the 1970s, cashing in on oil discoveries in

Getty at his English manor home, Sutton Place. He bought the 400-year-old mansion from the Duke of Sutherland.

the North Sea. During this time, he became a symbol of Americans' distrust of corporations when reporters revealed that he used his financial expertise to avoid paying income tax.

Jean Paul Getty remained president of Getty Oil until his death of prostate cancer in 1976 at the age of 83.

Legacy

Getty's primary legacy was as a pioneer in the technique of strategically purchasing stock in order to gain control of companies that were many times the size of his own operations. He also left $660 million of his estate to an art museum in Los Angeles that carries his name. Beyond that, he left little that has endured.

Getty was a man who worked best alone. He earned a fortune because he had no one to answer to.

He could be patient and accept losses for the hope of future earnings, whereas his competitors had to satisfy shareholders less likely to tolerate long-term gambles. But while Getty had no peer at making money, he knew nothing about creating anything of long-term value. A U.S. Navy report described him as "a financial genius at obtaining control but thereafter a genius at disorganization."

"A man is a business failure if he lets his family life interfere with his business record," Getty once said, but that attitude erased his name from the business world forever in a few short years. Because he paid little attention to family and associates, there was nothing to hold his vast empire together when he was gone. His family spent the decade following his death in bitter legal fights over the inheritance. Getty had been dead barely 10 years when Texaco absorbed Getty Oil for the price of $10 billion.

Getty sits alone at the head of his long dining table in Sutton Place.

7

ROBERT O. ANDERSON

ARCO'S NEW BREED OF OIL TYCOON

Imagine a shopper who goes to a store hoping to bring home the poorest quality merchandise available. If that sounds crazy, then you understand how bewildered Robert O. Anderson's business associates must have been when Anderson went shopping for a company to acquire. "I'm not looking for the *best* place to go, I'm looking for the *worst*," he said.

But there was method to this madness. Anderson had a gift for fixing struggling and failed companies. He knew that he could purchase these unattractive companies for a cheap price; then he could come in and make them profitable.

Such unconventional behavior was typical for Robert O. Anderson. He did not fit the mold of the humorless, tight-fisted, friendless oil baron

Robert O. Anderson (b. 1917) has been called "one of the last of the great wildcatters and oil tycoons of the twentieth century." In the course of his career, he balanced sharp business sense with environmental concerns.

whose life revolved entirely around making money. Although Anderson was a brilliant businessman who could figure numbers in his head faster than others could with a calculator, he was also a people person. Relaxed and outgoing, he made friends easily. Yet he also loved to spend time alone in the woods. Because of his appreciation for the outdoors, he sought to cooperate with environmental groups at a time when most oil tycoons saw them as mortal enemies.

Business was only one of many subjects that fascinated Anderson's curious mind. He originally planned for a career as a professor of philosophy. He generously supported the theater. An enthusiastically self-taught man, he once bet a friend that in one week he could learn enough about pottery to give an hour-long lecture on the subject with slides. He won the bet.

Growing Up in Chicago

The second of Hugo and Hilda Anderson's four children, Robert O. Anderson was born in Chicago on April 13, 1917. Hugo was a classic example of a man who started at the bottom and worked his way to the top. Originally an errand boy at the First National Bank of Chicago, he earned promotions all the way up to first executive vice president. Hugo Anderson was well respected for his honesty and for his wise loans to Texas and Oklahoma oilmen that helped the bank survive the lean years of the Depression. Most bankers considered these loans poor risks, but Hugo gambled on his ability to judge people's character. He seldom lost.

Self-conscious about his poor education, Hugo made sure his children grew up in the intellectual atmosphere of the neighborhood surrounding the University of Chicago. Since Hugo did not believe real estate was a good investment, the Andersons lived in apartments throughout Robert's youth.

Robert was so outgoing as a child that if his family sent him into a store to ask for directions, his brothers would have to go in 15 minutes later to pull him away from the conversation. He did so well in school that he won a four-year scholarship to the University of Chicago. Although interested in business, geology, and astronomy, he concentrated on the study of philosophy.

A Career in Oil

Robert had always been fascinated by the stories his father would tell when he returned from his trips to Oklahoma and Texas. Attracted by the drama of prospectors making or losing fortunes based on their ability to predict the location of oil, he decided to check out the industry firsthand. During the summer following his sophomore year in college, he traveled to Texas to work on an oil pipeline crew. The experience convinced him that he wanted to try his luck in the oil business.

Anderson graduated from the University of Chicago on August 25, 1939, and married Barbara Phelps the same day. As part of his plan to make a career in oil, he took a job with the American Mineral Spirits Company to learn about refining.

After two years on the job, Anderson heard that the Bell Oil and Gas Company wanted to buy MALCO, a small refining operation in Artesia, New Mexico, and was looking for someone to supervise it. He and his brothers, Hugo Jr. and Donald, offered their services, as well as $150,000 (most of it supplied by their father) to purchase a 50 percent share of the 35-employee refinery. Bell agreed and sent Robert out to run the place. On New Year's Day 1942, Anderson moved to the hot, desolate wasteland with his wife and baby daughter.

This MALCO refinery was built in 1931 by the Maljamar Oil and Gas Company and sold to Anderson in 1942. Although he sold the refinery in 1946, he later bought another refinery in Artesia and renamed it MALCO.

Efficiency Expert

Anderson immediately found ways to cut costs, and he purchased new equipment to increase the plant's efficiency. Within six months, daily production at the refinery jumped from 1,500 barrels to 4,000.

By this time, the United States had entered World War II and its military needed fuel. Anderson learned that an army air base near the MALCO refinery was so desperate for aviation fuel that it hauled it in from a source nearly 1,000 miles away. Given the high cost of transporting fuel, Anderson knew that his nearby refinery should be able to deliver the product at a fraction of the cost.

But when Anderson proposed adding aviation fuel to the refinery's products, his engineers told him that wasn't possible. Aviation fuel had to be of extra high quality, measured by an octane rating of 91. The best octane rating the refinery had produced prior to Anderson's arrival was 60. Anderson, however, refused to pass up the opportunity. He brought in equipment and changed procedures. Within 30 days, the plant raised the octane level to meet the air base's needs.

octane rating: a system used to measure how smoothly a fuel burns in an engine. Fuels with high octane numbers burn more smoothly than those with lower numbers.

Anderson accomplished so much in just a few months that, at the age of 24, he was made president of MALCO. He bought a second refinery later that year in Roswell, New Mexico.

Wildcatting

Now that he was established in refining, Anderson looked for a chance to break into production, which seemed the more exciting end of the business. He did not have the money to compete with large companies and so, as he had done with his refinery, he started small. Anderson bought some used drilling equipment for $1,400. Researching the area's geology on his own, he began to scout out promising places to drill, although he could afford only leases on land that had shown no indications of oil. After buying a lease near Roswell from Harry Steinberger, the old man he had hired to operate his equipment, Anderson began wildcat drilling in hopes of striking it rich.

Anderson's wildcatting instincts were not wildly successful. During his first few years, he drilled only one producing well. For a time, he debated buying some oil leases in northern Colorado, but he finally decided the price of $3 per acre was too high. Two years later, other investors found the largest oil field in the Rocky Mountains on that land. As he gained experience, however, Anderson developed a better knack for detecting oil. By 1945, he owned 20 producing oil wells.

In the late 1940s, Anderson capitalized on his ability to make sick companies well. Selling the now profitable Artesia refinery, he moved MALCO's operations to Roswell. He then bought another refinery in Graham, Texas, and improved its efficiency. Anderson moved into yet another facet of the oil business when MALCO launched its own pipeline company under the management of his brother Donald, who had recently returned from the war.

Personality Plus

Although these operations quietly made money for Robert Anderson, he was still a small-time operator. He remained virtually unknown in the oil business—until he pulled off a spectacular deal right under the noses of the industry giants.

Fred Manning owned one of the top 10 drilling companies in the world and held oil leases worth millions. In 1948, with his health failing, Manning began to entertain offers for his business. Undaunted by the competition from major executives who were courting Manning, Anderson entered the bidding. Instead of dealing with Manning's advisors, he called Manning directly on the telephone. Anderson's refreshing personality impressed Manning. The men entered into negotiations that lasted two years. Finally, in late 1950, they met in Denver for more discussion. Always able to think on his feet, Anderson listened to Manning's latest requests and then wrote out the terms of the contract on hotel stationery. For $11 million, Anderson bought some of the most coveted properties in the industry—450,000 acres of Manning's oil leases.

Anderson's friendliness helped him capture another choice piece of property a few years later. Wilshire, a large oil refinery south of Los Angeles, was having problems and was rumored to be up for sale. Unfortunately, the two chief owners of the refinery had developed a bitter feud. They so distrusted each other that if one agreed to terms of sale, the other would turn them down.

Anderson outhustled and outcharmed the major oil companies that were trying to purchase Wilshire. After getting both sides to trust him, he proposed terms to each with the assurance that he could get the other side to agree. Then he got Gulf Oil to put up 75 percent of the $8 million he needed for the sale in exchange for an agreement that MALCO would purchase a set amount of crude oil from Gulf. Anderson walked away with a contract that made him the fourth largest independent oil refiner in the United States, with an annual gross of more than $100 million.

"It wasn't a matter of price. That was a mistake some of the others made. It was a matter of psychology."
—Joe Lackey, Anderson employee, on the Wilshire Company deal

PATIENCE

While all this was taking place, Anderson patiently continued to increase his oil production. His main strategy was to wait for wells to be drilled on other leases near his own land. If oil was struck nearby, Anderson would come in and drill wells on his property. In 1942, for example, Anderson drilled on a piece of land and found nothing. He sat on that lease for nine years without doing anything. When oil was discovered on neighboring property, Anderson consulted his geological texts and then deepened his old well to 5,000 feet. Another duster.

The experts were wrong about Anderson's well because they arrived a day too late. At 5,000 feet, tests would have shown that the drill was poised on the very edge of the oil field that Anderson sought. Not wanting to waste money by standing idle, the crew drilled a few hundred feet further while they waited for the scientists. By doing so, they passed by the thin layer of oil, and the land the experts tested the next day was only barren rock. It would be six more years before Anderson, drilling in a different spot, found the huge oil pool.

Puzzled, Anderson brought in experts to determine if it was worth drilling further. The experts said no, but Anderson refused to give up. In 1957, 15 years after he had first drilled on the land, his advisors persuaded him to sell. Even then, Anderson held onto a few leases that he thought had the most promise. His stubbornness finally paid off later that

Anderson believed that "the petroleum industry, especially for the independent operator, is still the most exciting business a man can be in."

year when he struck a gigantic oil pool yielding over 250 million barrels on his remaining leased land.

Anderson's leadership style was often contradictory. He could be careful and calculating, with almost obsessive attention to the smallest details. For example, after purchasing Wilshire, he had its service station logo repainted three times until it was exactly the shade of blue he wanted. Yet at other times, he flew by the seat of his pants. When negotiating the sale of Wilshire to Gulf in 1957, he decided to ask for $25 million. At the last minute, on a whim, he upped his price to $26 million. Gulf accepted. Anderson joked that he had made a million dollars in a minute!

By the end of 1958, Anderson decided to concentrate totally on oil discovery and production. He sold all of his refining properties and reorganized his holdings into the Hondo Oil and Gas Company. This company did so well at locating new oil that by 1960, Anderson held at least a part ownership in more than 150 oil wells.

This success did not go unnoticed. The Atlantic Refining Company of Philadelphia, once part of the Standard Oil empire, desperately needed a cheap, ready supply of crude oil. One way to get it was to merge with a successful producer like Hondo. Although some experts wondered why he would want to get involved with a struggling company like Atlantic, Anderson saw weakness as an opportunity. Because of Atlantic's poor standing, the terms of the 1963 merger favored Anderson so that he ended up as the company's major shareholder.

Anderson had no intention of taking over the management of such a large company right away. But Atlantic turned out to be in such trouble that Anderson feared he would lose his huge investment if he didn't get involved. He agreed to take over as chief executive officer and set about fixing yet another broken company.

Anderson saw that Hondo alone could not fulfill Atlantic's need for new oil sources. The company continued to buy more than half of the crude oil it refined, often from its competitors. Whenever there was a price war, Atlantic's competitors would win by raising the price of the oil they sold to Atlantic. Anderson decided that Atlantic had to produce its own oil in order to survive. Since it did not have the financial resources to increase its own oil production, he looked for yet another merger.

At the same time, the California-based Richfield Company, a large oil exploration and production company, was under pressure to merge with an eastern refiner. In December 1965, it joined with Atlantic under Anderson's leadership. The Richfield employees were disgruntled at being taken over by a rival company, but Anderson won their confidence by giving the new organization a name that emphasized the equality of the partnership. In May 1966, the company officially became the Atlantic Richfield Company, or ARCO.

A Richfield gas station changes its sign to reflect the 1966 merger between the Atlantic and Richfield Companies, which formed ARCO.

Alaska

Before Anderson took over, the Richfield Company had been actively searching for oil in the wilderness of Alaska. Many companies believed there was considerable oil there. After all, the U.S. Navy had found oil during its exploration of the area known as the North Slope, where the northern Alaska coast descends to the Arctic Sea.

Drilling for oil in such a frigid climate, however, was enormously difficult and expensive. Equipment had to be brought into a remote area inaccessible by road or sea for most of the year. The ground was often frozen as hard as concrete, with a permanent layer of frost that penetrated as far as a thousand feet below the surface. No one had found any Alaskan oil worth extracting since Richfield's 1957 Swanson River discovery, the first and only commercial oil field in the state. Exxon and Royal Dutch/Shell had invested millions into Alaskan exploration without finding oil. Over the course of two years, British Petroleum and Sinclair had drilled six dry holes on the North Slope.

This long series of failures raised the distinct possibility that Richfield had wasted a great deal of money acquiring nearly a million acres of oil leases on the North Slope. But after visiting Alaska in February 1966, Anderson became convinced that a sea of oil lurked somewhere below the surface. Pooling its resources with the Humble Oil Company to defray expenses, ARCO started drilling 60 miles south of Prudhoe Bay on the north coast. The result, however, was more frustration. ARCO spent 10 months delving more than two miles into the ground without striking oil.

By this time, all the other oil companies started to shut down their drilling operations. Even British Petroleum, which had spent eight years in pursuit of Alaskan oil, abandoned the project. Anderson, though, refused to be discouraged. "If you can't take disappointment, you ought not to be in this business, since 90 percent of what you drill are failures," he said. Rather than pull out of Alaska to cut his losses, Anderson actually bought more leases.

In April 1967, Anderson moved his drilling rig, the only one remaining on the North Slope, up to Prudhoe Bay. The summer thaw, however, left the land a sodden mess. The crew had to wait until

ARCO's 1967 oil well amid the snow of Prudhoe Bay, Alaska

October when freezing temperatures firmed up the ground. Facing bitter winds and temperatures of -30 degrees Farenheit, they drilled to a depth of 7,000 feet. On December 26, the crew felt a violent vibration. Seconds later, a deafening blast of natural gas roared up from the ground. The crew continued drilling until they reached oil at 9,000 feet.

Although the discovery was encouraging, no one knew how large the oil field was. Anderson kept his fingers crossed as he awaited the results of a test drill seven miles away from the original. In the summer of 1968, this probe struck another part of the same oil field, confirming that it was the largest oil discovery in North America—a reserve estimated at 5 to 10 billion barrels.

Oil companies now faced the difficult task of bringing the Alaskan oil to refineries and markets in the continental United States and abroad. Pooling its resources with Humble and British Petroleum, ARCO proposed constructing a pipeline across Alaska from Prudhoe Bay on the Arctic Coast to the port of Valdez on the Pacific Coast. A total of eight companies eventually joined the effort, forming the Alyeska Pipeline Service Company. Before they could begin laying pipe, however, they had to build hundreds of miles of all-weather road to the oil field and design entirely new technology capable of pumping oil through 800 miles of frozen land.

While working out these problems, Anderson pulled off another astounding sales job. Sinclair, once one of the top 10 oil companies in the nation, had fallen on difficult times. The giant Gulf and

Before oil companies settled on a pipeline to transport Alaskan oil, they discussed some farfetched ideas—such as building a monorail across the state, developing nuclear submarine tankers that would carry oil under the polar ice cap to Greenland, or flying the oil in jumbo jets. They even considered laying down an eight-lane highway and having a fleet of trucks carry oil constantly across Alaska, until someone calculated that moving so much oil would require most of the trucks in America!

Western Corporation was attempting to absorb the company. Anderson moved into the picture, offering a merger with ARCO. In 1969, Anderson heard the final terms of the proposed agreement over a pay phone in the desert near his ranch. Without using a calculator or even a slip of paper, he computed every figure and approved the largest merger in the United States at that time.

Meanwhile, Anderson's Alaska triumph deteriorated into what he later called "the most frustrating five years of my life." In 1969, an oil well leak off the California coast polluted beaches and destroyed wildlife. This awakened a powerful environmental movement against oil companies and particularly against the trans-Alaska pipeline. Convinced that

In 1969, the first of 80 6,000-ton shipments of pipe arrived in Alaska from Japan and was neatly stacked in readiness. Due to controversy, work on the pipeline did not begin until five years later.

Environmentalists who opposed the Alaska pipeline were concerned that it would disrupt animal migration patterns. To address this problem, the Alyeska Company built 800 animal crossings along the pipeline, including 22 points at which the pipes—normally elevated several feet above the ground—ran underground so that herds of caribou and moose could avoid them. But some animals had no fear of the pipeline; in fact, they would stand under it for hours, enjoying the warmth radiated by pipes filled with hot oil.

the pipeline would destroy one the world's last unspoiled wildernesses, protesters fought the construction at every step. Squabbling among the oil companies produced further delays, as did legal challenges from Native Americans concerned about the impact of the oil production on their land. As a result, ARCO's equipment sat on the banks of the Yukon River for five years, waiting for action.

In October 1973, the Arab nations of Egypt and Syria launched a military attack on Israel. When America supported Israel in this conflict (known as the Yom Kippur War or the October War), other Arab nations retaliated by refusing to sell oil to the United States. This action, called an embargo, caused oil prices to rise over 370 percent and triggered a serious fuel shortage. It was the Arab oil embargo that finally broke the logjam in Alaska. Deciding that Alaskan oil was vital to the nation, Congress cleared all legal obstacles to the pipeline on November 16, 1973. Work finally began in mid-1974. By 1977, when the ARCO tanker *Juneau* sailed from Valdez with the first shipment of Alaskan oil, the total cost of the pipeline had reached almost $10 billion.

In 1982, Anderson signed an agreement to drill for oil in the South China Sea, making ARCO one of the first American companies to do business with Communist China. Later that year, he resigned as chief executive officer of ARCO. But he never lost his love of oil exploration nor the outdoors. By the time of his retirement, Anderson's investment on these two interests had made him the largest individual land owner in the United States.

The ARCO tanker Juneau

Legacy

Anderson's salesmanship and skill at putting together mergers helped to make ARCO America's sixth largest oil company, with revenues of nearly $13 billion in 1998. His persistence in Alaska led to the development of an oil field that, at its peak, supplied a quarter of all the oil produced in the United States.

The most lasting legacy of Anderson's career, however, is an awareness of the balance between industry and the environment. Prior to the 1970s, oil companies pursued oil wherever they could get it, without a great deal of attention to the impact they had upon the land. The clash over the Alaskan oil

In 1999, BP Amoco announced it would buy ARCO. When the Federal Trade Commission (FTC) challenged the $30 billion deal because it would create a monopoly on Alaskan oil, ARCO agreed to sell its holdings in Alaska to Phillips Petroleum. Even without these assets, the proposed merger will create the second-largest non-government oil company in the world.

The pipeline crosses three mountain ranges and 34 rivers and streams on its 800-mile journey across Alaska. Today, about 88,000 barrels of oil pass through the pipeline every hour.

that Anderson found focused national attention on the issue of resource protection.

Far more than most oil executives, Anderson welcomed efforts to forge a new balance between the acquisition of valuable resources and appreciation for the natural world. He was himself active in a number of environmental organizations. As he remarked of the Alaska experience, "We could no longer think of ourselves as a decent company producing decent oil and giving decent service. We were now deep into environmental affairs." Like it or not, all oil companies are in the same position.

The World's Oil Supply

Untapped reservoirs of oil and natural gas can still be found in many parts of the world, but only about one-third of this amount could be brought to the surface with existing technology. In the late 1990s, the world's oil reserves were estimated at 1 trillion barrels. Around 67 percent of the reserves were in Saudi Arabia and other countries of the Middle East. About 12 percent were in South America and Mexico, 7 percent were in Europe, and 3 percent were in North America.

Will these petroleum reserves be enough to meet the world's future energy needs? Since the 1970s, when shortages of gasoline led to long lines at service stations, people have been asking this question with increasing urgency. Some studies show that if world oil consumption continues to rise as it has in the recent past, existing reserves would be gone by the early 2000s. Other experts say that new and better methods now being developed to recover petroleum could greatly increase the available supply. Reducing world consumption of oil would be another way to avoid a shortage. Many conservation plans have been proposed and some have been tried, but use of oil continues to increase in many parts of the world.

As the world's available supply of oil dwindles, new technology and innovative drilling strategies are increasingly important. In 1996, Royal Dutch/Shell launched the Anasuria, *a state-of-the-art floating production storage and offshore facility, to develop oil fields in the North Sea.*

H
A
B
E
F
C
D
G
WORLD MAP OF MAJOR OIL DISCOVERIES
A: Titusville, Pennsylvania, 1859
B: Lima, Ohio, 1885
C: Corsicana, Texas, 1893
D: Beaumont, Texas, 1901
E: Bartlesville, Oklahoma, 1901-1905
F: Los Angeles area, California, 1920s
G: Lake Maracaibo, Venezuela, 1922
H: Prudhoe Bay, Alaska, 1967

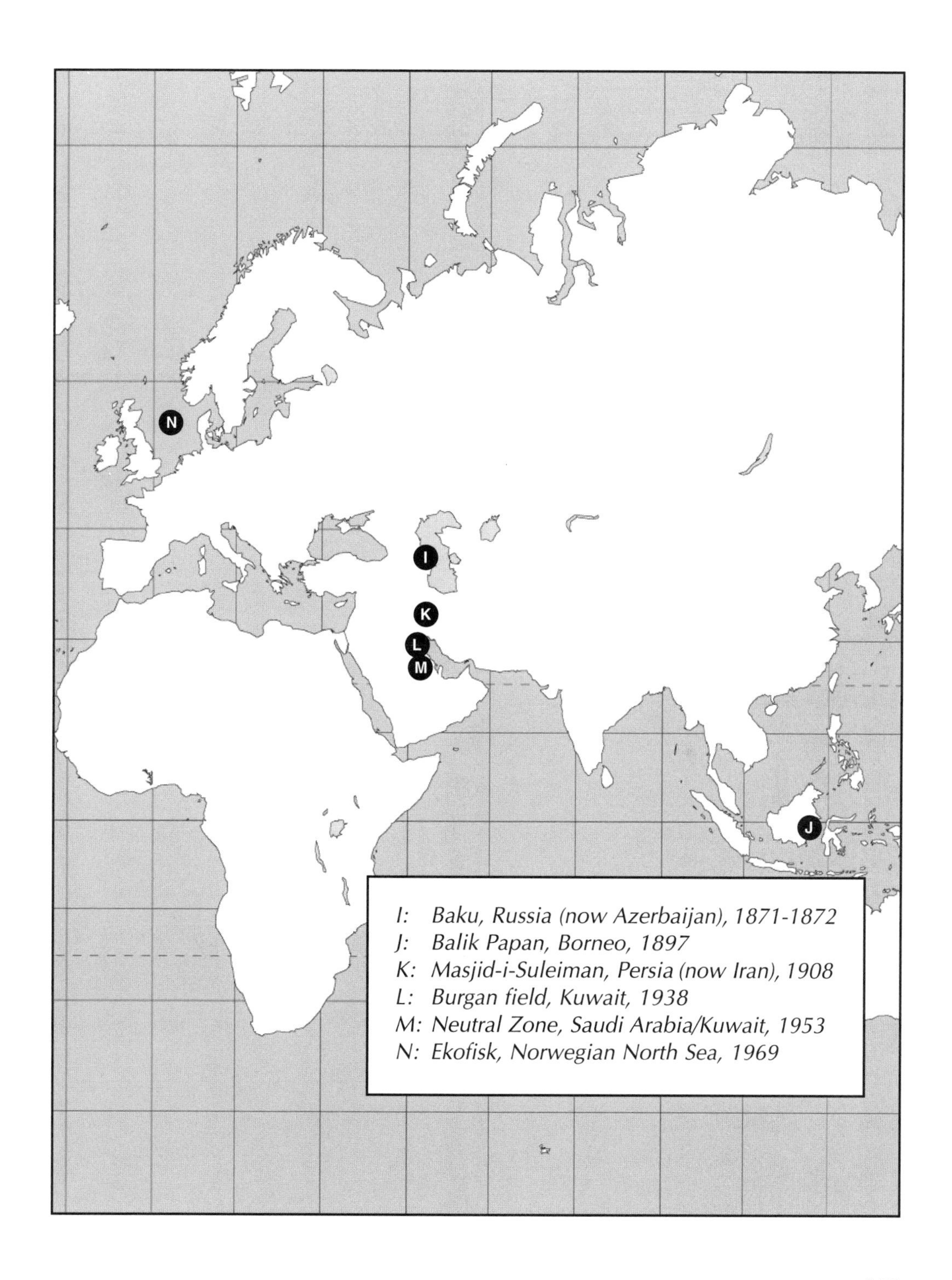
N
I
K
L
M
J
I: Baku, Russia (now Azerbaijan), 1871-1872
J: Balik Papan, Borneo, 1897
K: Masjid-i-Suleiman, Persia (now Iran), 1908
L: Burgan field, Kuwait, 1938
M: Neutral Zone, Saudi Arabia/Kuwait, 1953
N: Ekofisk, Norwegian North Sea, 1969

GLOSSARY

barrel: the standard unit of measurement for oil. One barrel equals 42 gallons.

bit: the cutting tool which, pushed deep into the ground by a drill, creates an oil well

cable-tool drill: an early drilling instrument. A heavy cutting tool was attached to a cable that lifted and dropped it repeatedly onto the ground to make a hole.

capping: restricting the flow of oil from a well. To prevent the oil from gushing high into the air, a system of valves is constructed at the point where the oil reaches the earth's surface.

concession: the right to search for oil and extract it in a particular area. Nations grant concessions to oil companies in exchange for payment of fees or royalties.

cracking: a refining process in which extreme heat and pressure are used to turn heavier fractions of oil into gasoline. Cracking increases the amount of gasoline that can be produced from oil.

crude oil: oil as it comes out of the ground before refining

derrick: the tower that supports drilling equipment over an oil well

drawback: an extra fee charged by railroads during the 1800s to transport oil for companies competing with Standard Oil. The railroads shared these fees with Standard, giving the oil giant an even greater advantage over its rivals.

drill: an instrument used to mine for oil by pushing a sharp tool deep into the ground, creating a well; *see also* **cable-tool drill** and **rotary drill**

duster: a well that produces little or no oil; also called a dry well

embargo: a government order that prohibits trade with another country, usually during wartime or for economic reasons

fraction: one of the various substances that make up oil. Each fraction condenses at a specific temperature, with "heavy" fractions condensing at higher temperatures and "light" fractions at lower ones.

fractioning: the process by which oil is separated into fractions. Oil is heated to a vapor and sent into a fractioning tower, where the fractions condense at different levels.

gasoline: a light fraction of oil, most commonly used as automobile fuel

gusher: an abundantly flowing well in which oil is forced to the surface by natural gas pressure

kerogen: a waxy substance made of fossilized plants and animals. Found in sedimentary rock, kerogen becomes oil when heated.

kerosene: a fuel produced from one of the light fractions of oil

lease: a contract granting the use of property for a specified time in exchange for payment; also, the property occupied under such a contract. The land on which a person or company drills for oil is often leased from private owners, who receive fees or royalties.

merger: an agreement that combines two or more corporations into one

monopoly: exclusive control over a product or service by one company, discouraging competition from other companies

natural gas: a naturally occurring gas found in the same reservoirs as oil and formed by the same forces of heat and pressure

octane rating: a system used to measure how smoothly a fuel burns in an engine. Fuels with high octane numbers burn more smoothly than those with lower numbers.

offshore drilling: drilling for oil that lies under a body of water. Most offshore drilling is done in seas or oceans, but oil deposits have also been found in large lakes.

petroleum: the scientific name for oil. Petroleum is a thick, flammable mixture of gaseous, liquid, and solid substances that forms naturally in rocks under the earth's surface. In Greek, "petroleum" means "rock oil."

pipeline: a long system of pipes through which oil is transported over land or underground. Pipelines can carry oil or natural gas from wells to refineries and from refineries to industries and homes.

pitch: a solid form of petroleum used for waterproofing and paving

refining: the process by which petroleum is treated, purified, and turned into finished products such as gasoline. A **refinery** is the industrial plant where this takes place.

reservoir: an underground deposit of oil or natural gas

rock oil: *see* **petroleum**

rotary drill: an instrument that mines for oil by rotating a cutting tool into the ground like a screw

rule of capture: the law that governed oil production in the early years of the industry. Landowners had the right to whatever oil they could bring to the surface on their land, even if the oil pool they drilled into lay below someone else's property or multiple properties.

salt dome: a cone-shaped formation of salt that pushes up through sedimentary rock, causing the rock to arch above it. Oil and natural gas may be trapped in reservoirs above or along the sides of salt domes.

sedimentary rock: rock formed from small particles of sand and mud

sediments: small particles of sand and mud that are changed into sedimentary rock by extreme heat or pressure

tanker: a ship in which oil is transported in large holding tanks

trust: a relationship in which a person or company holds the title to property for the benefit of another person or company. Trusts have sometimes been used, particularly at the turn of the twentieth century, to form large groups of companies that eliminate competitors by manipulating prices and carrying out other illegal business practices.

well: a deep hole drilled into the ground to obtain oil

wildcat: an oil well drilled in a place where no oil has been proven to exist

Workers construct a pipeline for Gulf Oil in Texas. The crews who labored in the early oil fields rarely achieved wealth or fame, but their contributions were invaluable to the growth of the oil industry.

Bibliography

Chernow, Ron. *Titan: The Life of John D. Rockefeller, Sr.* New York: Random House, 1998.

Clark, J. Stanley. *The Oil Century: From the Drake Well to the Conservation Era.* Norman, Okla.: University of Oklahoma Press, 1958.

Ferrier, R. W. *The History of the British Petroleum Company,* Vol. 1. Cambridge: Cambridge University Press, 1982.

Harris, Kenneth. *The Wildcatter: A Portrait of Robert O. Anderson.* New York: Weidenfeld & Nicolson, 1987.

Hawke, David Freeman. *John D.: The Founding Father of the Rockefellers.* New York: Harper & Row, 1980.

Henriques, Robert. *Bearsted: A Biography of Marcus Samuel.* New York: Viking, 1960.

Hoffman, William S. *Paul Mellon: Portrait of an Oil Baron.* Chicago: Follett, 1974.

Knowles, Ruth Sheldon. *The Greatest Gamblers: The Epic of American Oil Exploration.* New York: McGraw-Hill, 1959.

Lenzner, Robert. *The Great Getty: The Life and Loves of J. Paul Getty—Richest Man in the World.* New York: Crown, 1985.

Miller, Russell. *The House of Getty.* New York: Holt, 1985.

O'Connor, Richard. *The Oil Barons: Men of Greed and Grandeur.* Boston: Little, Brown, 1971.

Presley, James. *A Saga of Wealth: The Rise of the Texas Oilmen.* New York: G.P. Putnam's Sons, 1978.

Rundell, Walter, Jr. *Early Texas Oil: A Photographic History.* College Station, Tex.: Texas A & M Press, 1977.

Sampson, Anthony. *The Seven Sisters: The Great Oil Companies and the World They Shaped.* New York: Viking, 1975.

Wallis, Michael. *Oil Man: The Story of Frank Phillips and the Birth of Phillips Petroleum.* New York: Doubleday, 1988.

Yergin, Daniel. *The Prize: The Epic Quest for Oil, Money, and Power.* New York: Simon & Schuster, 1991.

Source Notes

Quoted passages are noted by page and order of citation.

Introduction

p. 16: Daniel Yergin, *The Prize: The Epic Quest for Oil, Money, and Power* (New York: Simon & Schuster, 1991), 29.

p. 17: J. Stanley Clark, *The Oil Century: From the Drake Well to the Conservation Era* (Norman, Okla.: University of Oklahoma Press, 1958), 33.

Chapter One

p. 20 (first): David Freeman Hawke, *John D.: The Founding Father of the Rockefellers* (New York: Harper & Row, 1980), 150.

p. 20 (second): Ruth Sheldon Knowles, *The Greatest Gamblers: The Epic of American Oil Exploration* (New York: McGraw-Hill, 1959), 11.

p. 21: Hawke, *John D.*, 12.

p. 22: Knowles, *The Greatest Gamblers*, 10.

p. 23 (caption): Ron Chernow, *Titan: The Life of John D. Rockefeller, Sr.* (New York: Random House, 1998), 54.

p. 27: Yergin, *The Prize*, 35.

p. 29 (first): Hawke, *John D.*, 53.

p. 29 (second): Hawke, *John D.*, 69.

p. 30: Yergin, *The Prize*, 42.

p. 33: (caption): Yergin, *The Prize*, 38.

p. 36 (first): Hawke, *John D.*, 2.

p. 36 (second): Hawke, *John D.*, 219.

Chapter Two

p. 39 (caption): Yergin, *The Prize*, 114.

p. 42 (margin): Robert Henriques, *Bearsted: A Biography of Marcus Samuel* (New York: Viking, 1960), 53.

p. 46: Henriques, *Bearsted*, 89.

p. 52: Yergin, *The Prize*, 116.

p. 53: Yergin, *The Prize*, 125.

p. 55: Anthony Sampson, *The Seven Sisters: The Great Oil Companies and the World They Shaped* (New York: Viking, 1975), 12.

Chapter Three

p. 60: Richard O'Connor, *The Oil Barons: Men of Greed and Grandeur* (Boston: Little, Brown, 1971), 69.

p. 61: Knowles, *The Greatest Gamblers*, 28.

p. 63: James Presley, *A Saga of Wealth: The Rise of the Texas Oilmen* (New York: G.P. Putnam's Sons, 1978), 45.
p. 66 (caption): Knowles, *The Greatest Gamblers,* 38.
p. 67: O'Connor, *The Oil Barons,* 85.
p. 69 (caption): Yergin, *The Prize,* 88.
p. 69: Sampson, *The Seven Sisters,* 39.
p. 71: Yergin, *The Prize,* 94-95.
p. 73 (caption): Sampson, *The Seven Sisters,* 39.
p. 74: William S. Hoffman, *Paul Mellon: Portrait of an Oil Baron* (Chicago: Follett, 1974), 52.
p. 75: Yergin, *The Prize,* 92.

Chapter Four

p. 79 (caption): Yergin, *The Prize,* 137.
p. 81 (margin): Yergin, *The Prize,* 137.
p. 83: Yergin, *The Prize,* 140.
p. 84 (margin): Yergin, *The Prize,* 155.
p. 86: R. W. Ferrier, *The History of the British Petroleum Company,* Vol. 1 (Cambridge: Cambridge University Press, 1982), 88.
p. 87 (caption): Yergin, *The Prize,* 147.
p. 87: Sampson, *The Seven Sisters,* 55.
p. 89: Yergin, *The Prize,* 147.

Chapter Five

p. 95: Michael Wallis, *Oil Man: The Story of Frank Phillips and the Birth of Phillips Petroleum* (New York: Doubleday, 1988), 17.
p. 97: Wallis, *Oil Man,* 48.
p. 98: Wallis, *Oil Man,* 67.
p. 100 (margin): Wallis, *Oil Man,* 358.
p. 101 (first): Wallis, *Oil Man,* 127.
p. 101 (second): Wallis, *Oil Man,* 128.

Chapter Six

p. 109: Robert Lenzner, *The Great Getty: The Life and Loves of J. Paul Getty—Richest Man in the World* (New York: Crown, 1985), 50.
p. 110 (margin): Lenzner, *The Great Getty,* xvi.
pp. 114-115: Yergin, *The Prize,* 439.
p. 118 (margin): Lenzner, *The Great Getty,* 196.
p. 120 (margin): Russell Miller, *The House of Getty* (New York: Holt, 1985), 181.
p. 121 (caption): Miller, *The House of Getty,* 278.
p. 123 (first): Lenzner, *The Great Getty,* 83.
p. 123 (second): Yergin, *The Prize,* 440.

Chapter Seven

p. 125 (caption): Yergin, *The Prize*, 570.

p. 125: Kenneth Harris, *The Wildcatter: A Portrait of Robert O. Anderson* (New York: Weidenfeld & Nicolson, 1987), 53.

p. 132 (margin): Harris, *The Wildcatter*, 40.

p. 133 (caption): Harris, *The Wildcatter*, 171.

p. 137: Yergin, *The Prize*, 571.

p. 139: Harris, *The Wildcatter*, 92.

p. 142: Harris, *The Wildcatter*, 97.

A fire rages through the crowded derricks and storage facilities of the Spindletop, Texas, oil field in the early 1900s.

Index

About the Author

Nathan Aaseng is an award-winning author of over 100 fiction and nonfiction books for young readers. He writes on subjects ranging from science and technology to business, government, politics, and law. Aaseng's books for The Oliver Press include the **Business Builders** series and nine titles in the **Great Decisions** series. He lives with his wife, Linda, and their four children in Eau Claire, Wisconsin.

Photo Credits

American Petroleum Institute: pp. 2, 55
AP/Wide World: pp. 133, 139
Archive Photos: pp. 108, 117, 119, 123
ARCO Photography Collection: pp. 124, 135, 137, 141, 142
Artesia Historical Museum and Art Center: p. 128
Bartlesville Area History Museum: p. 112
BP Amoco p.l.c.: pp. 76, 79, 80, 82, 83, 86, 87, 88
Chevron U.S.A. Inc.: pp. 65, 69, 75, 148
Dictionary of American Portraits (published by Dover Publications, Inc., 1967)**:** pp. 33, 73
Express Newspapers/Archive Photos: pp. 121, 122
Exxon (courtesy American Petroleum Institute)**:** pp. 13, 14
Library of Congress: cover (top right), pp. 18, 56 (bottom), 115
New York *Herald* (courtesy American Petroleum Institute)**:** p. 32
Pennsylvania Historical Commission, Drake Well Museum, Titusville, PA: pp. 6, 9 (courtesy American Petroleum Institute), 10, 16, 17, back cover
Phillips Petroleum Company: cover (top left), pp. 92, 95, 96, 97, 99, 101, 103, 104, 105, 107, 111
PhotoDisc, Inc.: cover (background)
Rockefeller Archive Center: pp. 21, 22, 23, 28, 37
Shell Photographic Library: cover (bottom right), pp. 38, 40, 43, 47, 48, 50, 51, 53, 143
Spindletop/Gladys City Boomtown Museum: cover (bottom left), pp. 56 (top), 60, 62, 64, 66, 68
Standard Oil of Ohio (courtesy American Petroleum Institute)**:** p. 26
Texaco Inc.: pp. 70, 71, 152
U.S. Signal Corps (courtesy Franklin D. Roosevelt Library)**:** p. 90